HERBAL
REMEDIES
HANDBOOK

MORE THAN 140 PLANT PROFILES
REMEDIES FOR OVER 50 COMMON CONDITIONS

ANDREW CHEVALLIER

THIS EDITION

Editorial assistants	Rosamund Cox, Poppy Blakiston-Houston
Senior editor	Kathryn Meeker
Senior art editors	Anne Fisher, Glenda Fisher
Senior producer, pre-production	Tony Phipps
Senior producer	Stephanie McConnell
Creative technical support	Sonia Charbonnier
Senior jacket designer	Nicola Powling
Managing editor	Stephanie Farrow
Managing art editor	Christine Keilty
Senior DTP designer, Delhi	Pushpak Tyagi
Pre-production manager, Delhi	Sunil Sharma

This edition published in 2018
First published in Great Britain in 2007 by
Dorling Kindersley Limited
80 Strand, London WC2R 0RL

Copyright © 2007, 2018 Dorling Kindersley Limited
A Penguin Random House Company
Text copyright © 2007, 2018 Andrew Chevallier
10 9 8 7 6 5 4 3 2 1
001–310356–Aug/2018

Important notice: Do not try to self-diagnose or attempt
self-treatment for serious long-term problems without first
consulting a qualified medical herbalist or doctor. Do not
take any herb without first checking the cautions in the
relevant herb entry and the guidelines in Safety and Quality
pp.40–51. Do not exceed dosages recommended. Always
consult a professional if symptoms persist. If taking
prescribed medicines, seek professional advice before
using herbal remedies.

A CIP catalogue record of this book
is available from the British Library
ISBN 978-0-2413-4202-2

Printed and bound in China

A WORLD OF IDEAS:
SEE ALL THERE IS TO KNOW

www.dk.com

WHAT IS HERBAL MEDICINE?

CONTENTS

FOREWORD

The last decade has seen a rapid increase in the publication of books on herbal medicine. However, it is rare to find one, such as this, that incorporates details of over-the-counter remedies, written for the general public by a practising herbal practitioner. Andrew Chevallier brings to this book years of experience in clinical practice and book authorship (he is the author of the *Encyclopedia of Herbal Medicine* – also published by Dorling Kindersley). Andrew is a well-known and respected member of the herbal fraternity in the UK, having been in the past President of the National Institute of Medical Herbalists.

The origins of the use of herbs dates back to prehistoric times, and their traditional medicinal use continues in all cultural groups throughout the world. Many modern medicines were inspired by constituents found in traditional medicinal plants, and some modern drugs are still isolated from plant materials. However, the big difference between herbal and modern medicine is that while modern medicines comprise single chemicals, nature has endowed each herb with a spectrum of active components, which work synergistically to produce a healing effect that cannot be replicated from single components.

Over the last twenty years there has been a resurgence of interest in the use of herbal medicine. There are several reasons for this, but the main ones are an increasing realization of the limits of modern medicine, particularly in the treatment of chronic disease; fear of adverse side effects of prescription drugs, and the increasing support for the medicinal use of plants from modern clinical research. However, herbal medicine need not be an alternative to modern medicine. Indeed, most herbs are mild medicines that work well alongside prescription drugs. Despite media "hype", true herb–drug interactions are few and these are clearly indicated, as appropriate, in this book.

Although the treatment of chronic disease always requires professional advice, with minor health complaints there is a lot you can do for yourself using over-the-counter herbal preparations. The key to finding successful natural remedies is knowing what you are doing and why. This book provides the necessary guidance. Eating a healthy diet is fundamental to maintaining good health, but herbs have a special role to play when minor health problems arise.

ELDERFLOWER

DR ANN WALKER
PHD MNIMH MCPP RNUTR
October 2006

Medicinal lore
Knowledge of medicinal plants often comes down to us from long-standing traditional use. Wild indigo root (*Baptisia tinctoria*) was used by Native Americans to bathe cuts and wounds and treat rattlesnake bites.

INTRODUCTION

Over the last two decades, herbal medicine has been flowering as people have rediscovered its many benefits. Herbs such as echinacea, lavender, and turmeric have become familiar self-help remedies, while the range of herbs and herbal products available online and in stores continues to grow.

Several hundred remedies, such as teas, tablets, or tinctures are easily found in health stores and pharmacies, while thousands can be purchased online. Finding safe and effective herbal medicines for home treatment may never have been simpler. And whether they are taken to treat common health problems, prevent illness, or enhance performance, herbal remedies – when used carefully – will work to improve health and relieve illness.

While choice provides opportunity, it often comes with a sting in its tail! Walking down the aisles of a health food store, past row upon row of competing products, can be a bewildering experience. If you want a remedy for sinus congestion or period pains, how do you make your choice and decide what to buy? If you know that you want elderflower (*Sambucus nigra*) for sinus congestion, or white peony (*Paeonia lactiflora*) for period pains, many questions still arise. How will the herb work? What is the best way to take it? Are there safety issues? What is good value for money? Is a single herb or a combination best? Is there a risk of interaction with any prescribed medication you're taking? Answering these questions in detail is beyond the scope of any one book, but *Herbal Remedies Handbook* aims to provide you with the essential information to answer such questions, to choose safe and appropriate herbal remedies, and to put them to good use.

Unlike synthetic medicines, herbal remedies are harvested from fields and forests the world over. Their healing, therapeutic activity is simply one aspect of the plant world's bountiful and generous relationship with animate life on our planet. With its many photographs and illustrations, this book reveals just a little of the beauty and variety to be found in medicinal plants, and encourages a greater awareness of the need to protect and conserve them in the wild. It is written in the hope that it will open up the world of herbal medicine for you and enable you to use herbal remedies to good effect.

Herbal stores
Herbal remedies are available all over the world and in many different forms. It pays to buy from stores that specialize in herbal medicine, with well-trained staff who can give sound advice.

WHAT IS HERBAL MEDICINE?

WHY USE HERBAL REMEDIES?

When confronted by illness and poor health, human beings have always sought medicines from the natural world, mostly from plants. Today, we have the opportunity to combine the insights of traditional herbal medicine with the clarity and quality control that scientific research can provide.

Back to nature

Strange as it may seem to us today, one reason why herbal medicines fell out of favour among doctors and scientists in the 19th and 20th centuries was because they were natural! By the middle of the 20th century in the West, conventional medicines, such as aspirin and antibiotics, had almost entirely replaced the old traditional herbal remedies. Though aspirin itself was originally based on a compound found in willow bark (*Salix alba*), it was now being produced from petroleum. At that time, it was possible to think that herbal medicines had been consigned to history and superseded by pharmaceutical medicines.

Yet over the past 50 years, the use of over the counter herbal medicines has grown and grown – research is 2013 found that more than 35 per cent of the UK population use herbal remedies each year. There are many reasons for this turnaround, but one key factor is that herbal remedies are by and large very safe and work to support healthy function rather than override it. "First, do no harm", as Hippocrates put it.

Herbal remedies are very suitable for home use and you do not have to be seeking to go "back to nature" to use them – though, this might help! Over the last decade, scientific research has discovered the key importance of our gut (and other) bacteria in maintaining good health. An unhealthy gut flora can lie behind digestive problems such as bloating and irritable bowel, as well as

more serious problems such as arthritis, diabetes, and psoriasis. It can even cause depression. In this sense, many foods and herbs, such as garlic (*Allium sativum*) and marigold (*Calendula officinalis*), both of which promote healthy gut and digestive function, support an ecological balance within the body – a "back to nature" within the body itself.

WHY USE HERBS?

Some of the strengths of herbal medicines are listed below, and you will find examples of them throughout this book.

- Along with the rest of life on our planet, human beings have evolved alongside plants, using them as food and medicine.
- Herbal remedies usually work with the body's own physiological processes, helping to strengthen underlying areas of weakness e.g., recurrent infection.
- When used sensibly, herbal remedies have an enviable safety record.
- They can safely be self-administered in minor acute and chronic health conditions.
- Many remedies can be taken long term to prevent illness or promote performance.

Lavender
Herbal medicines such as lavender (*Lavandula officinalis*) are natural products that offer a wide range of potential health benefits.

- Many are gentle acting and make good medicine for the very young and very old.
- Research has endorsed the use of a range of key herbal medicines, where both safety and effectiveness have been established. Other remedies have long-standing traditional use as evidence of their efficacy.
- As natural products, they are a permanently renewable resource.

The herbal medicine chest
Medicinal plants can be processed in many different ways. The commonest examples are teas, tinctures, and extracts made into capsules and tablets.

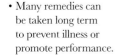

HERBAL HEALTH

This section looks at the broad picture of the way in which herbal remedies may be employed. For detailed advice on the self-treatment of particular problems, please refer to the Common Health Problems section on pp.256–275.

Herbal remedies can be safely used to:
• Treat common acute problems, for example coughs, headaches, and skin rashes.
• Treat chronic problems, for example mild depression, arthritis, and varicose veins.
• Prevent illness.
• Enhance health.

Though they are natural, herbal remedies are medicines and can therefore cause side effects. For best results, they need to be used sensibly and with respect. They also need to be used with an awareness of what they cannot do!

MINOR ACUTE PROBLEMS

Herbal remedies are well suited to treating everyday health problems, though the standard caution about self, or home, treatment always applies: if you are in doubt, seek immediate professional advice, especially where unwell children are concerned. Relief for conditions such as headache, sore throat, cough, wind, and bloating can occur quickly, though gradual, ongoing improvement in symptoms is more common with herbal medicine. Warm teas and diluted tinctures can be

particularly helpful. For simple problems, treatment for a few days will be sufficient. Skin problems such as minor burns, grazes, and rashes can be treated topically.

Example Symptoms of sore throat and hoarseness suggesting the start of a viral infection can be treated using remedies such as echinacea (*Echinacea* spp.), liquorice (*Glycyrrhiza glabra*), and sage (*Salvia officinalis*). Take tea or diluted tincture of one or more of these remedies, first as a gargle, then swallow. Continue taking three times a day until symptoms have cleared. If symptoms deteriorate sharply or there is no improvement after five days, seek professional advice.

CHRONIC PROBLEMS

Problems that have lingered for months or even years, such as acid indigestion, osteoarthritis, and fungal skin infections, can be successfully relieved or improved with herbal remedies, though long-term treatment may be required (a rule of thumb used by herbal practitioners is one month's treatment for every year the condition has been present). Taking a remedy regularly is likely to prove more effective than occasional dosing when symptoms flare. That being said, there is nothing wrong in using herbal remedies for symptomatic relief.

Example Acid indigestion can be relieved (and hopefully reversed) by taking meadowsweet

Applying sage
Fresh sage (*Salvia officinalis*) leaves are antiseptic and can be rubbed on insect bites and stings.

(*Filipendula ulmaria*) tea or tincture after meals for several months. Other remedies such as slippery elm (*Ulmus rubra*) and chamomile (*Chamomilla recutita*) may also prove helpful, as will attention to diet.

PREVENTING ILLNESS

Taken long term, many herbal remedies have been shown to have a potent ability to prevent illness or deterioration in existing symptoms. Indeed, some would argue that working on the basis that prevention is the best medicine is the way to use herbal remedies. The difficulty with a preventative approach of course is that, if successful, one does not see results. Such an approach nevertheless comes close to the ancient Chinese emperor's practice of paying his doctors only so long as he remained well. As a result it was in the doctors' interest to act preventatively rather than wait until the illness had taken hold.

Example Millions of people take ginkgo (*Ginkgo biloba*) on a daily basis in order to maintain healthy blood flow to the brain and limbs and to support memory and concentration, both uses that are endorsed by clinical research. Ginkgo also helps to protect nerve tissue and to counter allergic conditions.

ENHANCING HEALTH

Herbs can prove extremely helpful in promoting mental and physical performance, particularly where constitutional weaknesses exist. Whether taken by students sitting for exams or athletes preparing for an event (who should remember that herbal remedies may test positive in drug testing), remedies with tonic and adaptogenic properties support endurance and the ability to cope with stress. Night shift workers, people putting in long hours or tolerating extreme conditions in the workplace, and those suffering from long-term stress may all benefit from such remedies.

Meadowsweet
Meadowsweet tea or tincture is commonly taken to relieve digestive problems such as acidity, indigestion, and diarrhoea.

Example For exams and interviews, mental focus and vitality can be enhanced with herbs such as rosemary (*Rosmarinus officinalis*) and schisandra (*Schisandra sinensis*) – but note that when taking these herbs it is important to try them out first, before the exam or interview. Similarly, people working long hours or nights can benefit from remedies such Siberian ginseng (*Eleutherococcus senticosus*) or golden root (*Rhodiola rosea*) to improve their stamina and work rate.

Schisandra
A Chinese tonic herb, schisandra is commonly taken to improve liver metabolism and enhance mental stamina and performance.

ARE HERBAL REMEDIES SAFE?

With a few exceptions, all the herbal remedies in this book are recognized as safe for home use. The few that are not, such as lobelia (*Lobelia inflata*), are commonly included in manufactured herbal products and are safe when taken as instructed.

All the remedies listed have some evidence of effectiveness, though this varies greatly from plant to plant. The ability of ginger (*Zingiber officinale*) to relieve nausea and vomiting is strongly supported by research evidence. On the other hand, the use of chickweed (*Stellaria media*) to soothe itchy skin and eczema has barely been researched, and rests upon traditional knowledge and direct experience.

The safety of herbal remedies, and their effectiveness as medicines, is not necessarily easy to establish. By and large, knowledge about how herbs work comes from:
• Use as food or a food supplement.
• Traditional knowledge of use as a medicine.
• The experience of herbal practitioners.
• Scientific research.

NUTRITION

Remedies such as garlic (*Allium sativum*), lemon (*Citrus limon*), oats (*Avena sativa*), and soya (*Glycine max*) form a regular part of many people's diets, and are therefore used as both food and medicine. Long-term food use confirms that the remedy is safe to take as a medicine, though it says little about its effectiveness as a medicine. Some herbal remedies contain significant levels of nutrients and are taken as nutritional supplements, for example kelp (*Fucus vesiculosus*). Such use comes from scientific investigation of the plant's constituents.

KNOWLEDGE

The strength of traditional medicinal systems such as Ayurvedic (India/Sri Lanka), Chinese, and Western herbal medicine lies in the fact that knowledge and experience of remedies has been built up over thousands of years, constituting what has been described as the longest-ever clinical trial. Put to the test of time, it is argued, few herbs that are harmful or ineffective will remain in popular

Ouimum acrimi. ople. cala in £. fic.inp. Bceno bñ aloufez. uuuamichi.eï ïba ftingir. fic Laxat uentrem. Tcaunumm. obtenebrat uifu. Remotio.natumn.di potuliaca. Od gña- t humotc; acumm.a imflatuum. Juenir. fno. feibi.brcme a feptentonaliti:

An ancient tradition
Effective herbal remedies such as liquorice (*Glycyrrhiza glabra*) have been grown and used as medicines for thousands of years.

use – only those found to be safe and effective will retain their place. Long-standing use of a herbal remedy can therefore be seen as a strong indicator of safety and usefulness, though it is *not* a guarantee.

EXPERIENCE

Trained herbal practitioners develop a practical, subtle understanding of how best to apply herbal medicines and are watchful for signs of side effects. They are able to select those remedies most likely to help a patient. The collective experience of herbal practitioners – for example, caution in giving devil's claw (*Harpagophytum procumbens*) to patients with acid indigestion – can give important pointers to the safety and effectiveness of remedies.

RESEARCH

Scientific investigation into a medicinal plant spans a multitude of different types of research which add, like pieces in a jigsaw puzzle, to the overall picture. Researchers can investigate:

• The chemistry of the plant – its constituents and their actions.
• The whole plant – parts used, actions, uses, safety issues, and so forth.
• Processing – how to extract and process the remedy.
• Clinical trials – the therapeutic use of a plant extract, including dosage levels, safety, and evidence of effectiveness.

On one level, the chemistry of the plant, or phyto-chemistry, underpins all aspects of herbal research. If you know the key chemical constituents of a plant, you can make a reasonable guess about its level of safety and value as a medicine: caffeine is a key constituent in coffee (*Coffea arabica*), cola, guarana (*Paulinia cupana*), mate (*Ilex paraguariensis*), and tea (*Camellia sinensis*). Its stimulant activity forms part of the action of each plant.

Yet, each plant also has its own unique activity and character. The natural complex of constituents found within

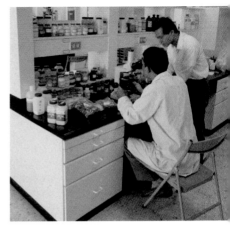

Modern methods
Scientific research starts by analysing the chemistry of the plant. Scientists can then investigate how key active components influence its medicinal activity.

a herbal remedy – the "whole" – is more than its key active constituents – "the sum of its parts". This interplay or synergy between different constituents is often a factor in the safety and effectiveness of a herbal medicine (see also Types of Herbal Remedy, pp.30–33).

Good clinical trials compare the safety and efficacy of a herbal medicine against another medicine or a placebo (a dummy product). Data collected from these trials usually provides the strongest evidence of just how safe or effective a herbal remedy is.

> ### NEW USE FOR CINNAMON
> The marriage between traditional knowledge and scientific research can lead to exciting new insights into the use of herbal remedies. Cinnamon (*Cinnamonum zeylanicum*) is a good example. While it is used in Asian traditional medicine for colds, flu, and digestive problems, recent investigations indicate that it has a potent stabilizing effect on blood sugar levels, helping to slow or prevent the onset of diabetes. It is also active against *Helicobacter pylori*, a bacterium commonly linked with stomach ulcers.
>
> **CINNAMON BARK**

HOW REMEDIES WORK

Around the world, people's knowledge of how herbal medicines work varies widely. In many traditional societies, spirits are thought to be responsible for a plant's activity, while in others the appearance and structure of a plant – its "signature" – indicates its use as a medicine.

ANIMAL MAGIC

Observation of how animals use medicinal plants has undoubtedly been a key factor behind the traditional use of many plants, and is now itself the subject of scientific study. Around the world, animals from bears to snakes have been observed eating plants with specific medicinal activity only at those times when they have need of them. Animals may also acquire a taste for plants that have general effects. The stimulant activity of coffee (*Coffea arabica*) was reputedly discovered after goatherds noted the frisky behaviour of goats feeding on the bush's red berries!

HOW PLANTS WORK

Many people, including herbal practitioners, believe that medicinal plants work in part on an energetic level, each plant having a distinct vitality, reflecting perhaps some kind of electro-magnetic force. Understanding the vitality, or vital force, of a remedy gives clues that can be used to refine its use as a medicine, in particular helping to match specific remedies with the needs of individual patients. That being said, scientific research provides the greatest certainty that a remedy is safe to use and has a reasonable chance of being effective. Scientific study of medicinal plants and their chemical constituents underpins our understanding of how herbal remedies work, and provides precise information on a plant's therapeutic activity and potential use as a medicine.

Nature's medicine
Bears are known to seek out and eat large quantities of antioxidant-rich roots and berries shortly before going into hibernation.

ACTIVE CONSTITUENTS

The main types of active constituent found in herbal remedies are listed below. Medicinal plants typically contain hundreds of different constituents with only a small proportion having direct therapeutic activity. As a rule, if you know a plant's active constituents you can broadly predict its medicinal effects, for example limeflower (*Tilia* spp.) contains a volatile oil with sedative activity, flavonoids, mucilage, and phenols. These constituents correlate with limeflower's standard use as a remedy: to aid sleep and relaxation, relieve headache and fever, and lower blood pressure and support the circulation.

CONSTITUENT	COMMON MEDICINAL ACTIVITY	EXAMPLES
Phenols	Often have anti-inflammatory, antiseptic, and antioxidant properties	Salicylic acid, found in willow bark (*Salix alba*)
Volatile oils	Complex mixtures of plant compounds with a wide range of actions, including stimulant, sedative, anti-inflammatory, and insecticidal properties	Essential oil of tea tree (*Melaleuca alternifolia*)
Flavonoids	Often pigments with purple, yellow, or white colour; many are strongly antioxidant and benefit the circulation, some are oestrogenic	Rutin, found in lemon (*Citrus limon*) pith and peel
Tannins	Have astringent, binding (or tanning) action; often with potent antioxidant and anti-inflammatory properties	Catechins, found in witch hazel (*Hamamelis virginiana*)
Coumarins	Often have blood-thinning or antispasmodic properties	Aesculin, found in horse chestnut seed (*Aesculus hippocastanum*)
Saponins	Key medicinal compounds similar in structure to the body's own hormones, often having hormonal or anti-inflammatory activity	Dioscin, found in Mexican wild yam (*Dioscorea villosa*)
Anthraquinones	Constituents that at the right dosage act as laxatives	Sennosides, found in senna (*Cassia* spp.)
Cardiac glycosides	Powerful compounds that act on the heart; often toxic	Digitoxin, found in foxglove (*Digitalis purpurea*)
Cyanogenic glycosides	Compounds that contain cyanide; at low doses valuable as sedatives and relaxants	Sambunigrin, found in elder leaves (*Sambucus nigra*)
Polysaccharides	Large molecules that typically have a demulcent/soothing effect on mucous membranes	Mucilage, found in slippery elm (*Ulmus rubra*)
Bitters	Strongly bitter-tasting compounds that stimulate appetite and digestive function and slow the heart	Amarogentin, found in gentian (*Gentiana lutea*)
Alkaloids	A diverse group of compounds, some with very potent activity as medicines, for example morphine	Isoquinoline alkaloids, found in Californian poppy (*Eschscholzia californica*)

LEMON

HORSE CHESTNUT

Medicinal plants also contain nutrients, vitamins, and minerals. In a few cases, for example alfalfa (*Medicago sativa*) and kelp (*Fucus* spp.), vitamin and mineral levels are significant, though in most cases trace levels only are present.

FOXGLOVE

AROUND THE WORLD

Looked at from a global perspective, herbal medicine is humanity's most important resource for treating and relieving illness. Though conventional biochemical medicine provides the bulk of medical treatment in Western countries, this is far from the case elsewhere.

In China and India in particular, traditional medicine is as popular as its conventional counterpart, and the vast majority of remedies used are herbal. In China, people choose whether to receive traditional or biochemical treatment, although in practice biochemical medicine is recommended for acute and life-threatening illness, and herbal medicine for chronic, long-term illness.

RESEARCHING THE HERBS

The sheer scale of traditional Chinese medicine means that research and development in China has brought new insights into herbal remedies. Clinical research in Shanghai into sweet wormwood (*Artemisia annua*) led to the discovery that it was a highly effective treatment for malaria. In 2015, the key researcher, Dr Tu Youyou, was awarded the Nobel Prize for medicine for her life-long study of sweet wormwood.

In many parts of the world, particularly Africa, the vast majority of medicines are herbal. In Ghana, over 80 per cent of medicines used are herbal, most being prepared from native West African medicinal plants. Paralleling developments elsewhere in the world, an African Pharmacopoeia was published on 2010 providing key scientific and medical data on 51 African plants, including *Withania somnifera*.

Even in Western countries, a sizeable minority of medicines used are herbal. About 20 per cent of the medicines used in the British National Health Service are herbal in origin, while in Germany 90 per cent of doctors routinely prescribe

Chinese medicine
Traditional Chinese medicine centres and herb suppliers are now a common sight in shopping centres the world over. In China, herbal remedies remain the preferred form of self-treatment for everyday health problems such as stomach upsets.

herbal medicines to their patients, hawthorn (*Crataegus* spp.) and saw palmetto (*Serenoa serrulata*) being common examples.

Germany is a world centre for research into herbal or phyto-medicines, and German research from the 1940s onwards has been responsible for establishing the safety and effectiveness of many popular over-the-counter remedies, for example echinacea (*Echinacea* spp.). Worldwide, research into herbal medicines is expanding at an unprecedented rate. Many countries have set up and support national research centres.

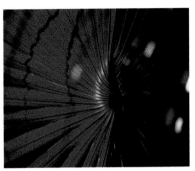

Saw palmetto
In Europe, saw palmetto is one of several herbal medicines routinely prescribed by doctors to treat symptoms resulting from an enlarged prostate.

OVER THE COUNTER

Sales of over-the-counter herbal remedies in the USA grew by nearly 50 per cent over the past 6 years. The top 10 individual bestsellers in the USA in 2016 are listed in the box shown below. Most of these individual remedies are also global bestsellers, though products combining several different herbal remedies are becoming increasingly popular.

Legislation in the European Union and in Australasia has established quality standards and labelling requirements for herbal remedies that should guarantee the quality of most products sold in health food stores and pharmacies. Online sales are much more difficult to police in terms of quality and safety. The legislation may have less impact upon internet or mail-order sales. In the USA, herbal remedies are generally classified as food supplements and are marketed under regulations governing food rather than medicines.

TOP 10 OVER-THE-COUNTER REMEDIES

The following 10 herbs were bestsellers in the United States in 2016:

1. Turmeric (*Curcuma longa*)
2. Wheatgrass / Barley (*Triticum aestivum / Hordeum vulgare*)
3. Flaxseed (*Linum usitatissimum*)
4. Aloe vera (*Aloe vera*)
5. Elderberry (*Sambucus nigra*)
6. Milk thistle (*Silybum marianum*)
7. Maca (*Lepidium meyenii*)
8. Withania (*Withania somnifera*)
9. Echinacea (*Echinacea* spp.)
10. Saw palmetto (*Serenoa repens*)

SOURCE HerbalGram 2017 (www.herbalgram.com)

ECHINACEA

HERBAL PRACTITIONERS

As herbal medicine has grown in popularity so has the need for trained practitioners who are able to assess a patient's needs and provide sound advice and treatment. In Western countries such as Australia, Ireland, the UK, and the USA, university training for herbal practitioners and naturopaths has become the norm; for these degree courses, practitioners are trained in both medical and herbal sciences. In China and India, universities teaching traditional medicine train practitioners to a standard equivalent to that of conventional medical practitioners.

USING HERBAL REMEDIES

MAKING HERBAL MEDICINES

The journey from hedgerow, garden, or medicinal plant farm to the finished herbal product takes many different forms. Wherever they are grown, however, medicinal plants need to be harvested and processed appropriately to achieve good-quality results and effective remedies.

Wild-crafting and cultivation

Many medicinal plants are still regularly picked from the wild – a process known as wild-crafting. Even in the developed world, herbs such as elder flowers and berries (*Sambucus nigra*), found in hedgerows and waysides throughout Europe, are wild-crafted both on a commercial basis, and locally, to make herbal medicines and medicinal wines.

In the developing world, herbs are as often wild-crafted as cultivated. In some cases, for example in some African countries, nearly 90 per cent of herbal

medicines used are gathered from the wild. Such a dependence on wild-crafting can threaten the survival of important medicinal plant species, especially if roots or bark are the main part of the plant used.

However, the main threat comes from commercial wild-crafting, where plants are gathered as a cash crop for export rather than for use locally as medicines. There are many examples of medicinal plants being pushed close to the point of extinction in this way – golden seal (*Hydrastis canadensis*) in North America and arnica (*Arnica montana*) in Europe are threatened species that are now being extensively cultivated. Until recently,

Endangered species
The survival of many medicinal plants is threatened in the wild. Buy organic or conservation grade products wherever possible.

echinacea (*Echinacea* spp.) was a common wild plant in its native North America; due to excessive wild-crafting it is now rare to find it in the wild.

The Convention on International Trade in Endangered Species of Wild Flora and Fauna (CITES) accord helps to prevent trade in endangered plant species, and by and large the needs of economics and conservation point the same way; cultivation makes better sense. Common medicinal plants such as German chamomile (*Chamomilla recutita*) are grown on a large scale in places as far afield as Egypt and Argentina. Demand for ginkgo (*Ginkgo biloba*) means that large plantations are now found in countries such as France and the USA, the leaves harvested by agricultural machinery. As demand for herbal medicine grows, so large-scale cultivation is more economically viable.

Organically grown medicinal plants are to be preferred. Being produced without chemical interference they are better placed to develop naturally and absorb nutrients from the soil. They should also be relatively free from inorganic fertilizers, pesticides, and pollutants. Organic certification provides some evidence that a plant meets certain quality standards and has been cultivated or wild-crafted in an ecologically sensitive manner.

Harvesting and drying

Whatever the size of the crop, the same basic rules apply. Though drying sheds and dehumidifiers are used industrially, a warm, well-ventilated drying rack such as an airing cupboard or a low-heated oven with the door open will suffice. Collect only plant material that you will be able to use or dry at once.

HARVESTING HERBS
- Try to harvest on a sunny, dry morning after the dew has evaporated.
- Ensure that you are picking the right plant, and the right part of it; using the wrong part may be dangerous.
- Use a sharp knife or scissors; cut perennials so as to encourage regrowth.
- Do not pick plants with blight or insect damage, nor plants growing in a polluted area.
- Plants are generally best harvested in the following stages: flowers, when just opening; leaves, when fully open; fruits, berries, and seeds, when ripe; whole plants, when mature.

DRYING
- Drying is best done in a shaded, well-ventilated area. Racks within a drying frame or airing cupboard are good, especially for leaves, flowers, roots, and bark, but whole plants can be hung up from a shaded line. Fresh plant material can be chopped and laid out on brown paper to dry – do not use newsprint as the inks are toxic.
- Discard poorly dried plant material, for example where the leaves are discoloured or where they show signs of fungal infection.
- Once dry, chop or break up the herb material into small pieces suitable for storage in labelled, sterilized glass jars or brown paper bags.

Dried herbs
Careful, unrushed drying, no matter how large or small the scale, is a key element in the production of good-quality herb material.

Industrial processing

Though many people's image of herbal medicine involves pans of bubbling liquids and strange-smelling bottles, the truth is that in many parts of the world, herbal products – tablets and capsules especially – are likely to be produced in a setting as far away from the kitchen worktop as it is possible to go. A large part of the herbal medicine industry is perhaps more accurately known as the phyto-pharmaceutical industry, with multi-million pound, high-tech, hermetically sealed factories. Many larger herb growers and phyto-pharmaceutical manufacturers are in fact owned by mainstream pharmaceutical companies. Enter one of the large phyto-pharmaceutical manufacturing sites and you will only find evidence of the herbal origin of its products in the Goods Inward section!

Remedies produced on an industrial scale to good manufacturing practice (GMP) standards are likely to be of good quality, and might be expected to be better than those produced in a more traditional, low-tech manner. However, much as is the case with small-scale, high-quality wine production, small-scale herb manufacturers often manage to produce plant material and herbal remedies of a distinctly higher quality than industrially produced over-the-counter remedies. That being said, GMP gives the buyer assurance that over-the-counter products should consistently meet acceptable quality standards.

CAPSULES AND TABLETS

Most capsules and tablets are industrially produced. Some are made with finely powdered herb material, though many are manufactured from soft or dry extracts. Various liquid solvents are used to dissolve the main active constituents

GINKGO TABLETS

found in the plant. These are then evaporated off, leaving a soft extract that contains about 20–30 per cent water. Dry extracts are easily powdered and typically contain no more than 5 per cent water. These extracts are then prepared, for example as tablets, to provide a uniform dose.

STANDARDIZED EXTRACTS

Standardized dried herb material or extracts are herbal products – typically capsules, tablets, and tinctures – which contain a minimum level of one or more key constituents. Using sensitive scientific measuring equipment, for example high-performance liquid chromatography (HPLC), batches of herb material are tested to establish levels of these constituents. Batches that meet the required levels of a given constituent are termed "standardized". Ginkgo (*Ginkgo biloba*), standardized on its flavone glycoside content, and milk thistle (*Silybum marianum*), standardized on its silymarin content, are common examples.

Some herbs are standardized on two different constituents: in the case of St John's wort (*Hypericum perforatum*), some products are standardized on its hypericin content, some on its hyperforin content, and some on both!

Most people would agree that this kind of quality control is valuable, especially in over-the-counter remedies. However, a more sophisticated method looks at the chemical fingerprint of the plant – the overall pattern of constituents extracted – that reflects the plant's natural complexity, sometimes referred to as a "full spectrum" of ingredients. Many people argue that this approach more accurately reflects herbal quality: it is the complex mix of constituents in a plant, not just one or two isolated constituents, that produces

Industrial production
Industrially produced herbal remedies made to GMP standards should be consistently good quality, and can be purchased with confidence over the counter.

its medicinal benefit. At least five different processes have been identified that contribute to the anti-depressant activity of St John's wort, for example. For this reason, herbal practitioners typically stock and dispense dried herbs and tinctures produced by small-scale manufacturers who use this "full spectrum" method of quality control.

PURIFIED EXTRACTS

A strong case can be made that standardized extracts that are purified – where levels of some constituents are enhanced at the expense of others – are not herbal remedies at all, but part-herbal, part-chemical medicines. Purified extracts are often highly concentrated – indeed, selected constituents can be so highly concentrated, by up to 2,000-fold, that few other plant chemicals will be present.

Such products may be valuable, but they cannot fairly claim to be herbal medicines. Red clover (*Trifolium pratense*) isoflavone extracts, and some aloe vera (*Aloe vera*) products, are examples of this type of process.

When buying herbal remedies, look at the label for information about standardization and refer to What the label should tell you on pp.50–51.

Skin creams
Good-quality herbal skin creams undergo the same quality-control procedures as capsules, tablets, or tinctures.

TYPES OF HERBAL REMEDY

Herbal remedies are prepared and used in many different ways, and it can often be hard to decide which form of remedy to select. The following section gives details of the main types of herbal preparation available, along with a summary of their pros and cons.

Herbal remedies can be made in a number of different ways, each method having its own specific advantages and disadvantages. Many preparations, for example teas and tinctures, are traditional and have been used to make medicines for thousands of years; no specialized equipment is required to make them. Others, however, require modern pharmaceutical methods of extraction and use a wide range of solvents and processes. Such extracts are most frequently made into tablets and capsules.

EXTRACTING THE ACTIVE

Though it is not commonly recognized, the processes used to make a medicine directly influence its effect on the body. For example, ginger root (*Zingiber officinale*) can be eaten fresh or dried in food, taken as a powder or capsule, made into a tea or tincture, or processed to produce a concentrated extract. In small but nonetheless

significant ways, each process extracts a different cross-section of chemicals within the plant – what is known as "the active". Ginger contains antiseptic resins that are poorly soluble in water, so ginger tea contains almost no resins; however, a ginger tincture made with 90 per cent alcohol (ethanol) extracts them well, thus having a greater antiseptic activity.

In day-to-day life, roughly the same applies to coffee. Espresso machines

COFFEE BEANS

were designed to extract maximum flavour and, contrary to popular belief, relatively low levels of caffeine. In contrast, coffee percolators extract far higher levels of caffeine while much of the finer flavour is destroyed by the continued heating. (Percolated coffee is not thought to be a healthy way to take coffee).

The following are the most frequently used types of herbal preparation. Details on making remedies are given in The home herbal on pp.36–37. For dosage advice, see pp.44–45.

Ginger
Fresh and dried ginger root are seen as distinct remedies in Chinese herbal medicine. As the root dries, new compounds are formed and some of the essential oil in the fresh root evaporates.

Herbal infusions
Brewing tea bags or an infuser in an open cup is fine for non-aromatic herbs. Brew herbs that contain essential oils in a teapot.

TEAS/INFUSIONS

Teas, or infusions, are the simplest way to make a herbal remedy, using the more delicate aerial (above-ground) parts of a plant – especially fresh or dried leaves and flowers. Teas are good for extracting water-soluble constituents such as flavonoids, for example in hawthorn leaves (*Crataegus* spp.), and essential oils, for example in peppermint (*Mentha* × *piperita*). Use a glass or ceramic (not metal) pot or cup with a lid to infuse the herb. Loose herb is generally better than teabags – it can be stirred and dispersed throughout the teapot or cup, improving extraction. Brew for 10 minutes then strain.

Pros Quick and easy to make; several herbs can be combined; fresh or dried herb material can be used; can be drunk, used as a mouthwash or gargle, applied as a lotion, hair rinse, and so forth; being diluted in water, fairly easily absorbed.

Cons Must be used quickly (maximum 24 hours, refrigerated); only water-soluble constituents extracted; taste can be unpleasant; a relatively large amount of liquid needs to be consumed.

DECOCTIONS

Decoctions are the most straightforward way to prepare tougher plant parts such as bark, berries, and roots. Chopped fresh or dried material is simmered in water for about 20 minutes. The resulting liquid is strained and drunk. Decoctions are good for extracting water-soluble constituents such as tannins, for example in yellow dock root (*Rumex crispus*).

Pros Can use fresh or dried herb material; several herbs can be combined; can be drunk, used as a mouthwash or gargle, applied as a lotion, and so forth; being diluted in water, fairly easily absorbed.

Cons Take a little time to make; must be consumed quickly (maximum 48 hours, refrigerated); taste can be very unpleasant.

DECOCTION OF SCHISANDRA

JUICES AND SMOOTHIES

These can be bought or made at home. Freshly prepared juices and smoothies can be an excellent way to take medicinal foods, e.g. beetroot juice or a "green" smoothie. Overall, smoothies are thought to be preferable as they retain the complex mix of constituents found in the food or herb. They support a healthy gut flora and aid digestive health. Use organic produce as far as possible.

Pros Relatively easy to make; 100 per cent natural product with high enzyme and micro-nutrient content; easily absorbed and aid digestive function.

Cons Must be kept refrigerated and, if bottled, consumed within 10 days; taste can be unpleasant.

TINCTURES

Tinctures are made by macerating (soaking) chopped herb material from any part of the plant in an alcohol solution, typically 45 per cent alcohol to 55 per cent water. The proportion of alcohol used varies from 25 per cent to 90 per cent, depending upon the active constituents to be extracted. Sometimes vinegar or glycerol is used instead of alcohol. Tinctures are relatively easily made and keep well, for three years or more. By using an alcohol and water mix both water-soluble and non-water- soluble constituents can be extracted, leading to a more concentrated product than is possible with teas or decoctions. The ratio of herb material to water and alcohol determines the strength of the tincture. An appropriate strength for most commonly available tinctures is 1 part herb material to 3 parts water and alcohol.

TINCTURE

Pros Long shelf life; different tinctures easily combined together; wide range of constituents extracted; small amounts effective; very easily absorbed.

Cons Takes several days to produce but can be easily purchased; can taste very unpleasant; contains alcohol (sometimes at high levels).

SYRUPS

Syrups are usually made by adding unrefined sugar or honey to infusions or decoctions at a ratio of 1:1 (half and half). Their sweetness can mask unpalatable herbs. The sugars help to soothe irritation within the throat and chest, and syrups are classically used as cough mixtures. Syrups, linctuses, and cordials can be bought over the counter or made at home.

Pros Fairly long shelf life; sweet-tasting and can be combined with unpleasant-tasting remedies; good for coughs.

Cons Large amounts of sugar; limited applications.

CAPSULES

Capsules generally contain dried powdered herb material or soft or dry concentrated extracts. Good-quality capsules are densely packed so that air cannot circulate through the powder. Vegetarian or non-vegetarian gelatine capsules (size 00) can be purchased for filling at home. Sealed capsules containing oils or concentrated soft or dry extracts are similar in action to tablets (see below) but usually contain fewer binding agents and additives.

GARLIC CAPSULES

Pros Convenient to take, and with little taste; clearly defined dose; often standardized.

Cons Cannot be blended like tinctures; powders can occasionally be irritant; may be excessively concentrated.

TABLETS

Tablets can be made by simply compressing dried herb material or a dry extract into tablet form, though usually herb material is combined with excipients (binding agents and additives) that maintain the tablet's shape and structure but dissolve in the stomach or intestines. Regrettably, artificial sweeteners and colours are often used in tablet formulation, so read the label.

Pros Convenient and concentrated; clearly defined dose; often standardized; little taste.

Cons May dissolve poorly in the digestive tract; may be excessively concentrated; cannot be blended like tinctures.

BOSWELLIA TABLETS

FIXED OILS

Unlike essential oils, fixed oils are made by soaking herb material in a vegetable oil, such as sunflower. Fixed oils are typically made with herbs that have wound-healing properties, for example marigold (*Calendula officinalis*), and can be applied neat to minor cuts, grazes, sprains, and so forth. They can also be used in creams and ointments.

Pros Easily massaged into the skin: blend well with essential oils.

Cons Often greasy – ointments or creams may be better.

FREEZE-DRIED EXTRACTS

Used increasingly in traditional Chinese medicine, freeze-dried extracts are made using a process similar to instant coffee. They have the full range of constituents found in the herb or herb formula.

Pros Convenient; clearly defined dose; often standardized.

Cons Expensive; not readily available.

POWDERS

Used traditionally in Ayurvedic medicine, powders are easy to take. They tend to deteriorate quicker than normal dried herb material.

Pros Easy to take.

Cons May taste unpleasant; need to be carefully stored.

MYRRH POWDER

OINTMENTS AND CREAMS

Ointments are made using oils and fats, and usually contain no water. Being oily they form a waterproof, protective surface on the skin and are most useful in conditions such as haemorrhoids and nappy rash. Creams are made by emulsifying oils and water in an emulsion, much as in mayonnaise. They are cooling and moisten the skin, and are used to soothe sore and inflamed skin conditions. Avoid applying ointments and creams to open wounds.

Pros Formulated for the skin.

Cons May contain artificial preservatives and stabilizers.

ESSENTIAL OILS

Essential oils are mostly produced by distilling flowers, leaves, and so forth and collecting the resulting oil – the plant's "essence". Being very concentrated, they must be used with care. Typically, they are blended in a carrier oil such as grapeseed oil at a *maximum* 5 per cent dilution, for example 1ml of essential oil in 20ml of carrier oil, or 5 drops in a teaspoon of carrier oil. Essential oils should not be taken internally, unless on the instruction of a suitably qualified doctor, medical herbalist, or naturopath.

LAVENDER ESSENTIAL OIL

Pros Highly concentrated; pleasant aroma; can be used diluted on skin or dispersed in air by burner.

Cons Can occasionally cause irritation or allergic skin reactions.

HERBAL REMEDIES AT HOME

Making simple herbal preparations is easy provided you follow a few straightforward rules – the most important being to start with correctly identified herb material. The dosages given below apply to most (but not all) commonly available herbal remedies – see pp.44–45.

PREPARING A HERBAL TEA

1 Add 1 heaped teaspoonful of fresh or 1 level teaspoonful of dried herb material (leaves, flowers) to a teapot. Pour in a cupful of water just off the boil.

2 Stir, cover, and leave the tea to stand for 10 minutes. Strain, add honey if wanted, and drink.

PREPARING A DECOCTION

1 Use 1 heaped teaspoonful of fresh or dried herb material (bark, berries, root) and put in a non-aluminium saucepan. Add 1½ cups of water and gently bring to the boil. Simmer for approximately 20 minutes.

2 Remove from the heat and carefully strain into a cup or other container. Add honey if wanted and drink. Larger quantities can be prepared if required, for example 25g (1oz) herb material to 500ml (18 fl oz) water.

PREPARING A TINCTURE

1 A standard tincture is made using 1 part dried herb material to 3 parts of alcohol solution. Place the chopped herb material – root, leaf, flower, or fruit – in a clean, preferably sterilized, glass jar or pan and stir in the required amount of alcohol. For each 100g (4oz) of dried herb material add 300ml (½pt) of alcohol solution (for strength see box shown right). Stir well, ensure the herb material is fully covered, close the lid, and label clearly.

2 Stir or shake the contents thoroughly for a few minutes each day, for 10 days. Strain into a sterilized glass bottle, seal with a cap, and label.

ALCOHOL SOLUTION

For tinctures, use organic vodka or other good-quality spirit of similar strength. Fresh plant materials require a 40 per cent alcohol solution. Dried plant material can be made with lower-strength alcohol, typically 25 per cent. Below 20 per cent, alcohol solution tinctures may decay.

STORING HERBAL MATERIAL AT HOME

DRIED HERB MATERIAL is best stored in glass jars. Jars should be clean and dry, and preferably sterilized. Fill the jar close to the top with herb material and close the lid firmly.

For short-term storage (a few months) dried herb material can be kept in clean brown paper bags, folded over and secured with a rubber band.

Label the jar or bag clearly with the following information:
• Name, and part, of plant
• Date of harvesting.

Keep in a cool, dry, dark place, away from heat and direct sunlight. Keep out of reach of children and animals.

Dried herbs should be used within 12 months of harvesting. Tinctures can remain effective for 3 years or more. Capsules and tablets should have a "Use By" date on the container.

Dried herb material that changes colour due to damp or fungal infection, or has mites (insect infestation), must not be used. Put it in the garden compost or place in a sealed plastic bag in a waste bin.

THE HOME HERBAL

Herbal remedies can prove highly effective in treating minor health and first aid problems. Cuts, grazes, headaches, digestive upsets, and so on can be dealt with effectively using just a few herbal products. Keeping a range of remedies at hand makes good sense.

First aid kit

Selecting herbal remedies to complement a standard first aid kit containing standard items such as plasters and scissors can be an enjoyable activity. The 10 remedies shown below are readily available in natural pharmacies and health food stores and are useful to have at hand for minor domestic emergencies. For details of their uses and applications, see the relevant entry in the A–Z section on pp.52–255.

Note Creams and ointments should not be put on open wounds: use water, aloe vera (*Aloe vera*), or witch hazel (*Hamamelis virginiana*) water to wash cuts and grazes, and then bathe with diluted echinacea (*Echinacea* spp.) or myrrh (*Commiphora molmol*) tincture. Warning! This will strongly disinfect the area but will sting painfully for a short time. Burns should be held under cold water for at least 10 minutes before applying aloe vera or lavender (*Lavendula* spp.) essential oil.

Echinacea capsules and tincture (pp.118-119)

Slippery elm powder or tablets (p.231)

Thyme dried herb and essential oil (p.221)

Myrrh tincture (p.107)

Aloe vera juice (pp.62–63)

Tea tree essential oil (p.163)

Comfrey ointment/ cream (p.212)

Lavender essential oil (pp.152–153)

Garlic, fresh or capsules or both (pp.60–61)

Arnica ointment/ cream (p.66)

Not all remedies have to come out of a medicine bottle or jar. The kitchen shelf contains some of the best remedies for home treatment. Use cleaned, fresh produce, avoid old or discoloured material, and ensure store cupboard items are within their use-by date. Here are some examples:

Barley – lemon barley water for urinary tract problems

Cinnamon – as a tea for colds, sore throat, and digestive disturbance

Cranberry – juice or powder for cystitis and gastro-intestinal infections

Olive oil – a few drops for wax in ears

Cayenne – a pinch or two in infusions or food for colds and flu

Clove – a clove or 1–2 drops of essential oil for toothache

Honey – applied topically as a dressing for small wounds and minor burns

Ginger – as a tea for nausea and indigestion, or with garlic for colds and flu

Cabbage – warmed leaf as a poultice for painful joints

HOME HERBAL PHARMACY

Any number of dried herbs or tinctures can be added to build up a home herbal pharmacy. The following are some of the more useful, versatile, and safe remedies:

Chamomile (*Chamomilla recutita*)
Cramp bark (*Viburnum opulus*)
Elderflower (*Sambucus nigra*)
Feverfew (*Tanacetum parthenium*)
Liquorice (*Glycyrrhiza glabra*)
Limeflower (*Tilia* spp.)
Meadowsweet (*Filipendula ulmaria*)
Peppermint (*Mentha* x *piperita*)
Rosemary (*Rosmarinus officinalis*)
Sage (*Salvia officinalis*)
Valerian (*Valeriana officinalis*)
Yarrow (*Achillea millefolium*)

THE HERB GARDEN

Growing herbs in the garden can bring real pleasure and a sense of achievement. Culinary herbs add zest to familiar recipes, and their scent lingers on summer evenings. At the same time, even a few medicinal herbs can become a first aid resource for common health problems.

Outdoor pharmacy

Beyond the kitchen shelf and bathroom medicine cabinet, your home herbal pharmacy can extend out onto the window sill, balcony, or garden. Given adequate light, water, and food, many medicinal plants will thrive and a surprising number can be grown even in a small area. Living medicinal plants are a useful resource to have at hand; they are also a joy to work with and to have in the garden. Label plants so that when you are harvesting them you know precisely what material you are gathering.

If you are planting in the garden, choose herbs such as lemon balm (*Melissa officinalis*) or yarrow (*Achillea millefolium*) that grow vigorously in most soils and

harvest well. Buying seeds or plants from specialist herb suppliers is usually the best way to stock a medicinal herb garden.

Other herbs that are readily cultivated at home include:
Calendula (*Calendula officinalis*)
Echinacea (*Echinacea* spp.)
Californian poppy (*Eschscholzia californica*)
Fennel (*Foeniculum vulgare*)
Lavender (*Lavandula* spp.)
Parsley (*Petroselinum crispum*)
Rosemary (*Rosmarinus officinalis*)
Sage (*Salvia officinalis*)
Thyme (*Thymus vulgaris*)
Feverfew (*Tanacetum parthenium*)
Vervain (*Verbena officinalis*)
Cranberry (*Vaccinium macrocarpon*)

CONTAINER HERBS

Many herbs will grow well in pots or other types of containers, including the following:

Lemon verbena (*Lippia citriodora*)
Peppermint (*Mentha x piperita*)
Oregano (*Origanum vulgare*)
Parsely (*Petroselinum crispum*)
Skullcap (*Scutellaria lateriflora*)
Feverfew (*Tanacetum parthenium*)
Heartsease (*Viola odorata*)
Basil (*Ocimum basilicum*)

THE RIGHT PLANT

Medicinal plant cultivation is fairly straightforward, but it is important to start with:

- The right plant species – *Aloe vera* is a safe and effective plant medicine, but some *Aloe* species are poisonous.
- The right variety – some varieties are more medicinally active than others. A specific variety of damiana (*Turnera diffusa* var. *aphrodisiaca*) is generally used.
- If the plant's essential oil is important, the right chemotype – plants of the same species and variety can differ in their chemical constituents, especially in their essential oils, for example lavender (*Lavandula angustifolia*).

Herb garden
In a small plot, select herbs that have culinary and medicinal use such as fennel (*Foeniculum vulgare*), sage (*Salvia officinalis*), and thyme (*Thymus vulgaris*).

SAFETY AND QUALITY

SAFETY CONCERNS

Herbal remedies are natural but they are also medicines, and can cause side effects. Like all medicines, they need to be treated with respect and used carefully. The following section gives information and simple advice on the risks associated with taking herbal remedies.

SIDE EFFECTS

Herbal remedies have an excellent track record when it comes to safety, and side effects are very infrequent. However, unwanted reactions do occur and it is important to be alert to this possibility, especially when taking a remedy for the first time. Adverse reactions to herbal remedies usually involve minor symptoms such as digestive upset and headache. On stopping the remedy, symptoms usually slowly clear. Sometimes, existing symptoms can flare up when starting a new remedy.

In either case, if you suspect that you are reacting badly to a herbal remedy, stop taking it. If symptoms are severe, or continue to worsen despite stopping the remedy, seek immediate advice from your health care practitioner. If the symptoms are minor, try an alternative remedy.

Remedies that contain potentially irritant or toxic constituents, for example horse chestnut (*Aesculus hippocastanum*), are more likely to produce side effects and need to be treated with caution. Sticking to recommended dosage levels is important, especially in the case of children. Taking excessive doses of any medicine is likely to lead to side effects, whether herbal or conventional.

Allergic reactions to herbal medicines, even familiar ones such as German chamomile (*Chamomilla recutita*), are rare but do sometimes occur. Mild allergic reactions should begin to ease soon after contact with the remedy is ended – applying marigold (*Calendula officinalis*) cream and drinking nettle (*Urtica dioica*) tea can help with minor skin reactions. Severe allergic reactions are a medical emergency and need immediate medical attention.

For people who have allergies to plants or foods, or are known to be sensitive to medicines, it's a good policy to start new herbal remedies by taking a small amount, say half the minimum recommended dose. If everything is fine, build up over a few days to the standard dose; if it's not fine, stop!

Known cautions are listed for each herb in the A–Z of Herbal Remedies, pp.52–255.

Half dose
If prone to allergies start with a low dose.

Dosage
Horse chestnut seed contains saponins that can irritate the gastrointestinal tract. At the normal dosage, side effects are unlikely to occur.

CONTRA-INDICATIONS

Some herbal remedies need to be avoided in pre-existing health conditions, as they may worsen symptoms. For

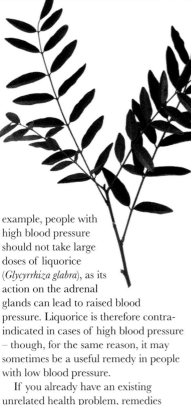

Liquorice
The root contains constituents that stimulate the release of hormones by the adrenal glands. This action is mainly responsible for its effectiveness as an anti-inflammatory remedy.

minor and go unnoticed, but in some cases herb–drug interactions can cause serious, even life-threatening problems.

The potential of St John's wort (*Hypericum perforatum*) to interact with prescribed drugs has been intensively researched, after it was found that it speeded up clearance of a number of drugs from the body. A report in *The Lancet* (2000) described the case of a heart transplant patient recovering well in hospital who went quickly downhill after taking St John's wort. On investigation it was discovered that St John's wort had caused levels of ciclosporin, an immunosuppressant drug, to drop by 50 per cent, leading to his body starting to reject his new heart. On ceasing to take St John's wort, he quickly recovered.

This is an extreme example, but it illustrates that herb–drug interactions are real and need to be taken into account. If you are taking drugs prescribed by your doctor or hospital, check with them, or with a registered herbal or naturopathic practitioner, before taking a herbal remedy.

If you experience an adverse reaction to a herbal remedy in the UK you can now report it at: yellowcard.mhra.gov.uk.

example, people with high blood pressure should not take large doses of liquorice (*Glycyrrhiza glabra*), as its action on the adrenal glands can lead to raised blood pressure. Liquorice is therefore contra-indicated in cases of high blood pressure – though, for the same reason, it may sometimes be a useful remedy in people with low blood pressure.

If you already have an existing unrelated health problem, remedies should be carefully selected in order to avoid using those that are contra-indicated. Each remedy listed in the A–Z of Herbal Remedies can be checked for known contra-indications.

Pregnancy and breast-feeding are the situations where herbal remedies are most commonly contra-indicated. Some remedies are unsuitable to take during pregnancy and breast-feeding, and in the first three months of pregnancy all medications including herbal remedies should be avoided as far as possible. For more details, see Pregnancy and after on pp.46–47.

HERB–DRUG INTERACTIONS

Some herbs (and foods) influence the effects of conventional medicines, interacting with them and increasing or decreasing their strength of action. Often interaction between a herb, for example schisandra (*Schisandra chinensis*) or cat's claw (*Uncaria tomentosa*) and a drug will be

---

MAIN HERB-DRUG INTERACTIONS

The main categories of prescribed drugs requiring caution are:

• Anticoagulants, antidepressants, anti-epileptics, and immuno-suppressants; the effectiveness of the contraceptive pill can also be affected.
• If you are taking prescribed medicines, do not stop taking them in order to start taking a herbal remedy. Seek professional advice from your doctor or herbal practitioner on the best way forward.

DOSAGE AND SENSIBLE USE

As with all medicines, getting the dosage right is essential. Too much and you risk overdosing, too little and the remedy may not work. Follow the guidelines on these pages to ensure that you use herbal remedies safely and appropriately.

ADULT DOSAGES

Each of the remedies listed in the A–Z of Herbal Remedies has a letter indicating its adult dosage – how much of the herb to take per day or per week.

To take an example, passion flower (*Passiflora incarnata*) on p.173 has C for its dosage. Looking at the dosage guide (*right*), it can be seen that C = 2–4g a day or 30g a week. Passion flower should therefore be taken at these recommended dosages.

As another example, hawthorn leaf (*Crataegus* spp.) has M and C for its dosage. M applies to manufactured products: take pre-packed hawthorn products, such as standardized tablets and capsules, at the manufacturer's recommended dosage. C applies to dried hawthorn leaf or berry: take at the recommended daily or weekly dosage, i.e. 2–4g a day or 30g a week.

Similarly, each of the other letters gives specific recommendations on how to use the herb.

Teas and decoctions The dosages given in the guide apply when making teas and decoctions from dried herb material – bark, leaves, roots, etc. For fresh herb material you can use 1½–2 times the quantity of dried material.

Tinctures It is not possible to give clear guidelines for tinctures owing to the wide variation in their strength. Ask advice on dosage when purchasing a tincture. In general, the dosage range for a 1:3 tincture is the same (in millilitres not grams) as the above dosages, i.e. for A, the dosage of a 1:3 tincture is 5–15ml a day.

ADULT DOSAGE GUIDE

Recommended ADULT dosage as given in the key information boxes (see opposite page). For children and the over 70s, see below and opposite.

A	=	5–15g a day, or max. 100g (3½ oz) per week
B	=	3–7.5g a day, or max. 50g (2 oz) per week
C	=	2–4g a day, or max. 30g (1 oz) per week
D	=	1–2g a day, or max 15g (½ oz) per week
M	=	Take product at manufacturer's recommended dosage
T	=	Topical application on the skin only (Note: preparations made specifically for topical use should not be taken internally.)

Powders Take the minimum recommended daily dosage only.

Tablets and capsules Take at the manufacturer's recommended dosage.

CHILDREN'S DOSAGES

Do not give babies under 6 months any medication without professional advice. You may need to adjust dosage levels for children who are particularly small or large for their age.

From 6 months to 1 year: give ¹/₁₀ the minimum adult dose. From 1 to 6 years: give ⅓ the minimum adult dose. From 7 to 11 years: give ½ the minimum adult dose. From 12 to 16 years: give the low adult dose.

DOSAGES FOR OVER 70s

As we age so our bodies become less efficient at breaking down drugs, including herbs. From the age of about 70 onwards it is advisable to take slightly lower doses: 80 per cent of the standard adult dose is normally recommended. In very old and frail people the dosage may need to be as low as 50 per cent of the standard adult dosage.

GENERAL CAUTIONS

- Do not take essential oils internally unless on advice of a suitably qualified health care professional.
- Do not give herbs to babies under 6 months old.
- Do not exceed the recommended dosage levels.

- If you are taking drugs prescribed by your doctor or hospital, check with them, or with a registered herbal or naturopathic practitioner, before taking a herbal remedy.
- People known to have allergies should start by taking a low dose and, if this is fine, then increase the dose.
- Contact allergy can occur on handling fresh or dried herbs. Where such allergy occurs, do not take the remedy internally. Some people are allergic to specific plant families, for example the daisy (*Asteraceae*) family. Several herbs listed in this book, including chamomile (*Chamomilla recutita*), echinacea (*Echinacea* spp.), and feverfew (*Tanacetum parthenium*), belong to this family and are known to cause contact allergy in sensitive individuals.

Key information

Every remedy in the A–Z features a key information box that provides essential data on the herb. At the top, each herb is rated using a 5-star rating system, with 5 black stars = most. This gives some idea of the herb's:

- overall safety record (Safety)
- long-standing use in traditional medicine (Traditional use)
- evidence of effectiveness, as supported by scientific research (Research).

On the line below (Best taken as), suitable types of preparation are recommended; for example, yarrow (*Achillea millefolium*) is best taken as a tea, which gets 3 ticks. Dosage information is provided on the following line. Some entries include an "Often used with" recommendation. The last and most important section lists known cautions for the remedy, and should be read carefully, especially before taking a remedy.

overall level of safety ———
use of herb supported by research ———
guidelines on dosage ———
common herbal combinations ———

KEY INFORMATION

SAFETY   ★ ★ ★ ★ ☆
TRADITIONAL USE ★ ★ ★ ★ ★
RESEARCH   ★ ★ ★ ★ ★
BEST TAKEN AS Fresh raw garlic ✓✓✓
Whole garlic extract/aged garlic (as tablet or capsule) ✓✓
DOSAGE Long-term: 1 clove a day; short-term: up to 3 cloves a day; manufactured products: M (*see pages 44–45*)
OFTEN USED WITH Echinacea species (*Echinacea*)
CAUTIONS If taking anti-coagulants (blood-thinning drugs) such as aspirin or warfarin, take garlic only on the advice of a herbal or medical practitioner. *See also pp.41–52.*

——— use in herbal medicine
——— the best form to take the herb
——— safety guidelines

PREGNANCY AND AFTER

Herbal medicine has an important role to play as part of a natural approach to health care for expectant mothers and their children. While caution is required in avoiding potentially harmful remedies, many gentle-acting herbs can be safely used during pregnancy and beyond.

Herbal medicines pre- and post-natally

Herbal remedies can be safe to take during pregnancy, though for the first three months of pregnancy, they should be taken only on professional advice. From the fourth month onwards, a range of safe and effective remedies may be used to treat simple health problems such as colds, catarrh, and constipation. Select remedies with a long history of use in pregnancy and with no evidence of risk to mother or baby. The box below gives examples.

Check labels of herbal products carefully, especially where herbs and other constituents are combined together. Alcohol should be avoided during the first three months, even in the small amounts present in tinctures. It is much better to use teas, decoctions, tablets, or capsules.

HERBS TO AVOID

Some of the herbs included in this book are contra-indicated and unsafe to take during pregnancy and while breast-feeding. In particular, do not take:

- Chiretta (*Andrographis paniculata*)
- Neem (*Azadirachta indica*)
- Golden seal (*Hydrastis canadensis*)
- Lobelia (*Lobelia inflata*)
- Butterbur (*Petasites hybridus*)
- Pau d'arco (*Tabebuia impetiginosa*)
- Coltsfoot (*Tussilago farafara*).

Essential oils, and herbs that contain strong essential oils such as eucalyptus (*Eucalyptus* spp.), thuja (*Thuja occidentalis*), and sage (*Salvia officinalis*), are also contraindicated and should not be used during pregnancy and while breast-feeding.

SOME HERBS COMMONLY USED IN PREGNANCY AND WHILE BREAST-FEEDING

When pregnant and while breast-feeding, check the relevant cautions before taking a remedy. Remember that herbs are passed on to the baby in breast milk.

Garlic (*Allium sativum*)
Calendula (*Calendula officinalis*)
Senna (*Cassia* spp.)
Chamomile (*Chamomilla recutita*)
Echinacea (*Echinacea* spp.)
* Raspberry leaf (*Rubus idaeus*)

Butcher's broom (*Ruscus aculeatus*)
Elderflower/berry (*Sambucus nigra*)
Limeflowers (*Tilia* spp.)
Nettle (*Urtica dioica*)
Slippery elm (*Ulmus rubra*)

Bilberry (*Vaccinium myrtillus*)
Cramp bark (*Viburnum opulus*)
Ginger (*Zingiber officinale*)
Cornsilk (*Zea mays*)

* Take raspberry leaf only in the last 3 months of pregnancy; see also p.191.

Herbal medicines and children

By and large children respond very well to herbal remedies, even if the taste can make administering them a bit of a struggle! Adding honey or mixing with apple juice will usually help to make remedies more palatable. Tablets or capsules can usually be opened and ground up, and taken on a spoon mixed with honey, maple syrup, and so forth.

Children typically fall ill and recover quickly. This can be very alarming for parents, as a healthy child at 8am can be a very unwell one by 11am. The main worry in acute illness is controlling fever and keeping the temperature below 39°C (102°F). Children with fevers approaching this level (and above) need medical attention.

That being said, the very unwell child at 2pm can be running around again by 6pm; children bounce back. Get to know your child's typical pattern when falling ill – you will often be able to recognize the difference between day-to-day problems and a potentially serious illness. The former can be safely treated with herbal remedies, while the latter needs professional advice and treatment. If in doubt, always err on the side of caution and seek advice.

> **SOME HERBAL REMEDIES SUITABLE FOR CHILDREN**
>
> Guidelines on dosage levels for children are given in Dosage and sensible use, pp.44–45.
>
> Garlic (*Allium sativum*)
> Marshmallow (*Althaea officinalis*)
> Caraway (*Carum carvi*)
> Chamomile (*Chamomilla recutita*)
> Echinacea (*Echinacea* spp.)
> Californian poppy (*Eschscholzia californica*)
> Cinnamon (*Cinnamomum verum*)
> Eyebright (*Euphrasia officinalis*)
> Meadowsweet (*Filipendula ulmaria*)
> Sea buckthorn (*Hippophae rhamnoides*)
> Elecampane (*Inula helenium*)
> Plantain (*Plantago* spp.)
> Blackcurrent (*Ribes nigra*)
> Yellow dock (*Rumex crispus*)
> Elderflower/berry (*Sambucus nigra*)
> Thyme (*Thymus vulgaris*)
> Limeflowers (*Tilia* spp.)
> Red clover (*Trifolium pratense*)
> Slippery elm (*Ulmus rubra*)
> Nettle (*Urtica dioica*)

Babies up to 6 months
Herbs are not safe for young babies, but breast-feeding mothers can take suitable remedies for them.

TIPS FOR HOME USE

Self-treatment of minor health problems makes sense, and besides helping one to feel better, can be very satisfying. These pages give a few tips on how to refine the use of herbal remedies. In cases of more serious illness, it is wise to consult a qualified herbal practitioner.

How to choose the right remedy

- Decide what the main symptoms are.
- Select remedies that are known to help these symptoms, for example cranberry (*Vaccinium macrocarpon*) for cystitis, elderflower (*Sambucus nigra*) for colds and sinus congestion.
- Develop experience in using specific remedies and build up your own stock of herbal remedies.

HOW MANY TO USE

- Combining 2–4 herbal remedies together can prove more effective, particularly if the problem is stubborn or recurs frequently; for example recurrent cystitis infection may be treated with cranberry, plus remedies such as buchu (*Barosma betulina*), echinacea (*Echinacea* spp.), and cornsilk (*Zea mays*).
- Combine remedies as teas or tinctures, or purchase a product containing the required remedies.

WHEN AND HOW MUCH TO TAKE

- Herbal remedies are generally best taken with water about 30 minutes before a meal.
- Take the recommended daily amount in 2–3 divided doses, ideally before your breakfast, lunch, and evening meal.
- Moderate to high doses of a herb can be taken for a few days for minor acute problems, for example if you are experiencing a sudden onset of cystitis symptoms, take cornsilk at the upper end of its dosage range (10–15g a day) for 3–4 days.
- Low to moderate doses should be used for long-term problems, for example

Cranberry juice
Cranberry is a classic home remedy for cystitis. Drink up to ¾ litre (1⅓pt) unsweetened juice a day for a few days to treat acute symptoms.

in the case of chronic bladder irritation, take cornsilk regularly at a lowish dosage (5–7.5g a day).

HOW LONG TO TAKE FOR

- Self-limiting conditions such as a sore throat or stomach upset that are safe to treat at home should clear up within 10–14 days at the most. If you are not fully recovered by then, see your health care practitioner.
- Start treatment as soon as possible, before symptoms become full-blown.
- Some remedies may be taken long term to prevent or treat chronic illness, for example ginkgo (*Ginkgo biloba*) to maintain healthy mental function or boswellia (*Boswellia serrata*) to provide relief for arthritic pain and stiffness.

When to seek professional advice

- Do not put off getting professional advice when symptoms are worrying, especially if severe pain or a temperature of over 39°C (102°F) is present, or if symptoms deteriorate sharply or unexpectedly. Professional advice, including telephone help lines, will help to ensure that you have the right treatment at the right time.
- More detailed guidance on when to seek professional advice is given in Common Health Problems, pp.256–275.

HOW CAN A HERBAL PRACTITIONER HELP?

- Herbal practitioners are trained, often at university, to use their in-depth knowledge of herbal medicine to assess and treat a wide range of health problems. They are able to give detailed advice on the best remedies and products to use, together with appropriate advice on diet and lifestyle. Where appropriate, they will refer to other practitioners, including GPs, and they can advise on interactions between herbs and drugs.
- If you have ongoing health problems, or want access to natural health care for your family, contact herbal practitioners in your area and find one that you

> **WHAT HEALTH PROBLEMS DO HERBALISTS TREAT BEST?**
>
> Herbal practitioners specialize in treating health problems with herbal medicine, and give advice on diet and supplements. The following types of conditions often benefit from treatment:
> - Allergies
> - Anxiety and stress-related problems
> - Arthritic and rheumatic conditions
> - Chronic infection and fatigue
> - Chronic inflammatory diseases
> - Mild to moderate depression
> - Digestive complaints
> - Menstrual and menopausal problems
> - Skin disorders

feel comfortable with and trust. Check that your practitioner, or medical herbalist, is insured and a member of a professional body such as the National Institute of Medical Herbalists in the UK. He or she should provide you with details of likely costs of initial and follow-up consultations and herbs.

Herbal dispensary
Medical herbalists stock a wide range of medicines. Each patient receives an individually tailored prescription, normally dispensed on the spot.

HOW TO BUY REMEDIES

When buying a remedy, select a suitable herb or combination of herbs for your health problem, such as feverfew (*Tanacetum parthenium*) for migraine headaches. Decide how you want to take it, for example as a capsule, and compare the different products available.

Where to obtain herbal remedies

Most people working with herbal medicine recommend buying remedies from a reputable health food shop or pharmacy. Try to find a shop that specializes in natural medicines, where the staff are knowledgeable and receive regular in-house training. Their advice can help to guide you towards buying appropriate, good-quality remedies. Shops that sell dried herbs and dispense tinctures are likely to have the most knowledgeable staff and should be actively concerned to supply good-quality herbal produce. Do not be afraid to ask how they can be sure that the products they sell are of good quality.

Buying remedies from a herbal practitioner is also a reliable way of obtaining them. He or she will stock products from growers and suppliers with a long-established record of quality control. Remember that the ethical code of a professional herbalist prevents him or her from selling you a herbal remedy without a consultation, for which you will usually have to pay. This is to assess your state of health and ensure that you receive the right advice and treatment for your situation. However, you will probably find that herbs supplied by a practitioner compare favourably in cost and quality with herbs obtained elsewhere.

Herbal remedies can, of course, also be bought online. In some cases you can source innovative products that are hard to find in a local health food shop. However, it can be very hard to distinguish between suppliers of genuine products and those that are there for the quick sale. Some internet and mail order companies deliberately base themselves in countries where proper quality control can be bypassed.

Where they are available, choosing organically certified herbal medicines makes sense on several counts, as plants are grown free of pesticides and pollutants, sites are regularly inspected, and the harvesting of organic herbs supports conservation in the wild.

QUALITY AND VALUE FOR MONEY

The price of herbal remedies is directly linked to quality. As a general rule, buy remedies produced by specialist herbal companies or well-known manufacturers, where effective quality control should be routine. Retailers who consistently market remedies at prices lower than the norm are either making a loss or selling products of doubtful quality! The best-value remedies are often sold at middle-of-the-road prices, reflecting acceptable quality. In some cases, for example ginseng (*Panax ginseng*), genuine high-quality products are available, and the higher price reflects this.

Sadly, poor-quality herbal products are still commonplace. If you take St John's wort (*Hypericum perforatum*) for mild to moderate depression, for example, it is reasonable to expect some signs of improvement. If no change occurs, this might be due to poor quality. Try a different product, perhaps a tea or tincture, instead.

BUYING DRIED HERBS

If you are purchasing dried plant material, bear in mind that herbs will soon deteriorate if they are not stored properly:

- Herbs should be stored in clear glass jars or brown paper bags, kept out of direct sunlight and away from damp or heat.
- Good-quality herbs should have their distinctive smell and taste – for example, marigold should be a vibrant orange colour, nettle leaf a deep dark green.
- Old or poorly dried herbs will be faded and will have lost their normal colour.
- Do not buy dried herbs in greater quantity than you need.
- For the pros and cons of different herbal preparations, see Types of herbal remedy on pp.30–33.

What the label should tell you

Different countries have different regulations governing what can (or cannot) be put on product labels. Nevertheless, any herbal product worth buying should have the information shown in the example below printed on the label or included as an insert:

In general, you should select herbal remedies that provide all this information; avoid buying products on which it is lacking. In particular, do not buy products where neither the local nor the botanical names of the herbs are given.

Checking the label
A reputable herbal product should give all the information shown in this example. When looking at similar products, compare the amount of herb material each provides. Also check what part of the plant has been used.

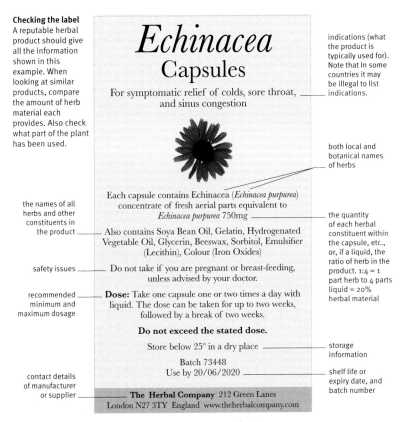

the names of all herbs and other constituents in the product

safety issues

recommended minimum and maximum dosage

contact details of manufacturer or supplier

indications (what the product is typically used for). Note that In some countries it may be illegal to list indications.

both local and botanical names of herbs

the quantity of each herbal constituent within the capsule, etc., or, if a liquid, the ratio of herb in the product. 1:4 = 1 part herb to 4 parts liquid = 20% herbal material

storage information

shelf life or expiry date, and batch number

Echinacea
Capsules
For symptomatic relief of colds, sore throat, and sinus congestion

Each capsule contains Echinacea (*Echinacea purpurea*) concentrate of fresh aerial parts equivalent to *Echinacea purpurea* 750mg

Also contains Soya Bean Oil, Gelatin, Hydrogenated Vegetable Oil, Glycerin, Beeswax, Sorbitol, Emulsifier (Lecithin), Colour (Iron Oxides)

Do not take if you are pregnant or breast-feeding, unless advised by your doctor.

Dose: Take one capsule one or two times a day with liquid. The dose can be taken for up to two weeks, followed by a break of two weeks.

Do not exceed the stated dose.

Store below 25° in a dry place

Batch 73448
Use by 20/06/2020

The Herbal Company 212 Green Lanes
London N27 3TY England www.theherbalcompany.com

A–Z OF HERBAL REMEDIES

YARROW

Achillea millefolium

Used as something of a cure-all, yarrow is an excellent remedy for colds, flu, and fever. It is equally good for healing cuts and bruises and for slowing or stopping bleeding, for example, nosebleed. The ancient Greek hero Achilles reputedly used yarrow on his wounded troops during the Trojan War.

MEDICINAL USES

Parts used Aerial parts (flowers)

Key actions Astringent ● Digestive tonic ● Stimulates sweating and reduces fever ● Stops bleeding ● Strengthens blood vessels (especially veins) ● Wound healer

Colds, flu, and fever Yarrow is most commonly taken as a tea to ease cold and flu symptoms, control associated fever, and speed recovery. Drink the tea hot, as this strongly stimulates sweating and encourages cooling. Increased sweating can help to reduce fever and leads to improved cleansing of waste products from the body. It combines particularly well with elderflower (*Sambucus nigra*).

FLOWERS   **DRIED AERIAL PARTS**

Other uses As a mild bitter, yarrow stimulates appetite and digestive activity, and is useful in treating diarrhoea and irritable bowel syndrome. It will help to reduce heavy menstrual bleeding and aids menstrual regularity. Yarrow is also a remedy for the circulation, helping to lower blood pressure, strengthen capillaries (small blood vessels), and tone varicose veins. Topically, yarrow ointment can be applied to heal bruises, sprains, and abrasions.

KEY INFORMATION	
SAFETY	★ ★ ★ ☆ ☆
TRADITIONAL USE	★ ★ ★ ★ ★
RESEARCH	★ ☆ ☆ ☆ ☆
BEST TAKEN AS	Tea ✓✓✓ Tincture ✓✓ Capsule ✓
DOSAGE	B (see pp.44–45)
OFTEN USED WITH	Elderflower (*Sambucus nigra*)
CAUTIONS	Allergic reactions can occur (especially skin irritation); not advisable during pregnancy and while breast-feeding; not advisable for children under 5. See also pp.42–51.

Yarrow in the Scottish Highlands was made into an ointment and was applied as a salve to heal wounds and bruises and treat skin problems.

CHIRETTA

Andrographis paniculata

Native to India, chiretta has a powerfully bitter taste that stimulates digestive and liver activity and counters infection. Highly valued in both Ayurvedic and Chinese medicine, its traditional uses include serious health problems such as diabetes, dysentery, fever, malaria, and worms.

MEDICINAL USES

Part used Whole plant

Key actions Anti-inflammatory
● Bitter tonic ● Immune-stimulant
● Protects liver

Poor immune system, liver and digestive problems A herb that looks set to become much better known in the West, chiretta has a therapeutic profile that is almost unique. On the one hand, it supports and strengthens the liver, protecting it from infection and toxic damage, while on the other, it has a marked ability to stimulate the body's immune system, making it more able to ward off and resist infection, especially viral infection. While the herb has been most used for liver problems in the past, its immune-enhancing properties make it a key remedy for protecting against upper respiratory infections, including colds, influenza, and tonsillitis. Several good-quality clinical trials have concluded that chiretta can reduce symptoms and improve the rate of recovery in people suffering from sinusitis, flu-like colds, and throat infection. It can also be helpful in gastrointestinal problems, such as food poisoning, gastroenteritis, and diarrhoea. Studies suggest that chiretta is as effective as paracetamol in relieving flu and fever symptoms, taken consistently over several days for best results. Chiretta is best taken on professional advice in liver disorders, and it works well with other liver remedies such as milk thistle (*Silybum marianum, see p.208*). Chiretta is also being investigated for its ability to help prevent cancer.

KEY INFORMATION

SAFETY	★ ★ ★ ★ ☆
TRADITIONAL USE	★ ★ ★ ★ ☆
RESEARCH	★ ★ ★ ☆ ☆

BEST TAKEN AS Tablet ✓✓✓ Capsule ✓✓ Tincture ✓

DOSAGE B: infusion, tincture. Andrographis extract (standardized to 4% andrographolides) as tablets: 1,020mg a day (*see pp.44–45*)

OFTEN USED WITH Echinacea (*Echinacea* spp.)

CAUTIONS Excess dosage may cause digestive discomfort; do not take during pregnancy and while breast-feeding. See also pp.42–51.

CAPSULES

While chiretta is used principally as an immune-enhancing remedy in the West, it is seen mainly as a digestive remedy in Ayurvedic and Chinese medicine.

HORSE CHESTNUT

Aesculus hippocastanum

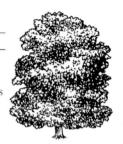

Originally from south-eastern Europe and Asia, the horse chestnut or "conker" tree is known around much of the world, though its health benefits are less well recognized. The shiny brown seeds – poisonous if eaten – are processed to make an effective medicine for the veins.

MEDICINAL USES

Part used Seed

Key actions Anti-inflammatory
- Astringent • Reduces fluid retention
- Vein tonic

Venous circulatory problems Horse chestnut is a major remedy for the veins and capillaries. As an astringent and anti-inflammatory, it has a beneficial effect on veins throughout the body, tightening up and toning the vein walls where they have become damaged and sore. By drawing back fluid that has leaked out of veins, horse chestnut reduces swelling and congestion in veins, as well as local inflammation, and is the first choice in herbal treatment for varicose veins and venous insufficiency (poor vein health). It is mostly taken in

seeds are a major remedy for varicose veins

seed, or conker

FRESH SEEDS

the form of a standardized tablet or capsule, although it may also be applied to the skin overlying varicose veins as a lotion, ointment, or gel. It should not be applied on broken or ulcerated skin. Its effectiveness has been fairly well established, though it usually needs to

The horse chestnut is found worldwide in temperate regions, and is widely grown in northern and western Europe an an ornamental tree.

HORSE SENSE

In Turkey, horse chestnut was used to treat chest complaints in horses, donkeys, and mules, and its common name may derive from this practice. As early as the 16th century, herbalists noted that "Turks call them horse chestnuts because they are very helpful for treating panting horses".

be taken for several months for signs of improvement, as healing venous circulation can be a very difficult task. A clinical trial at London's Barts Hospital in 1996 showed that horse chestnut extract was as effective as a compression stocking in treating varicose veins in the leg.

Other uses Horse chestnut may be used to treat other problems affecting the veins, for example, haemorrhoids and thread veins, and can be useful in treating leg cramps, and swelling and fluid retention in the legs. It may be taken to treat conditions such as deep vein thrombosis, frostbite, and leg ulcers, but only on professional advice. As a preventative, it will help to reduce the risks of deep vein thrombosis, for example on long haul flights. An oil extracted from the conkers has been used as a topical application for rheumatism in France, while in the USA, a decoction of the leaves has been considered useful for whooping cough.

KEY INFORMATION

SAFETY	★ ★ ★ ☆ ☆
TRADITIONAL USE	★ ★ ★ ★ ☆
RESEARCH	★ ★ ★ ★ ☆

BEST TAKEN AS Standardized extract (tablet) ✓✓✓
DOSAGE Tincture: C Tablet: 90–150mg of standardized extract (16–21% triterpene glycosides [aescin] a day) *(see pp.44–45)*
OFTEN USED WITH Butcher's broom *(Ruscus aculeatus)*
CAUTIONS Not suitable for children; may cause digestive irritation (not in enteric-coated preparations). If pregnant or breast-feeding, or taking blood-thinning medication, take on the advice of a herbal or medical practitioner. See also pp.42–51.

Although seeds are the most commonly used plant part, a decoction of the bark can be used in place of the seeds as an astringent lotion for varicose veins.

Cultivated as an ornamental tree in temperate regions of the world, the horse chestnut (*Aesculus hippocastanum*), has spiny green fruit with brown seeds, which are used medicinally.

GARLIC

Allium sativum

One of the world's most important medicinal plants, garlic is also one of the most researched, with over 1000 published papers investigating its therapeutic activity. Folklore has it that garlic protects against the devil and vampires, beliefs that attest to its power as a medicine, particularly in countering infection.

MEDICINAL USES

Part used Clove (one segment of the bulb)

Key actions Antibiotic ● Antifungal ● Blood-thinner ● Counters cough and respiratory infection ● Lowers blood pressure ● Lowers cholesterol levels ● Supports beneficial intestinal flora

Infections Before modern antibiotics became commonly available, garlic was one of the most frequently used remedies to treat infection. In the First World War, it was used to dress battle wounds. Although we have far more potent antibiotics today,

a tonic food and medicine

has antiseptic volatile oil

BULB

CLOVE

increase resistance to infection

GARLIC PEARLS

garlic still has a place in treating infection. It makes an excellent remedy for all types of respiratory infections, including sinusitis, cold, flu, sore throat, cough and, as a specific, bronchitis. Swallowed whole (one small clove), eaten crushed in with food, or taken as a tablet, garlic will strengthen the body's ability to fight infection and speed recovery. A simple and effective kitchen remedy for colds, sore throats, and coughs can be readily made by mixing a crushed clove of garlic with the juice of a freshly squeezed lemon (*Citrus limon*), 1–2 teaspoons of honey, and a pinch of dried ginger (*Zingiber officinale*) powder or, preferably, a small piece of chopped fresh ginger root. Place in a mug, add hot water, and stir. Drink up to three cups a day. Garlic may be taken alongside, and is likely to complement, prescribed antibiotics, at the same time warding off possible side effects by helping to protect beneficial intestinal bacteria and reducing the chances of developing thrush. Applied consistently over several weeks, fresh garlic or garlic oil may prove successful in countering local fungal infections such as itchy ear passages and warts.

KEY INFORMATION

SAFETY	★ ★ ★ ★ ☆
TRADITIONAL USE	★ ★ ★ ★ ★
RESEARCH	★ ★ ★ ★ ★

BEST TAKEN AS Fresh raw garlic ✓✓✓ Whole garlic or allicin-rich extract (as tablet or capsule) ✓✓

DOSAGE Long-term: 1 clove a day; short-term: up to 3 cloves a day; manufactured products: M *(see pp.44–45)*

OFTEN USED WITH Echinacea species (*Echinacea* spp.)

CAUTIONS If taking anticoagulants (blood-thinning drugs) such as aspirin or warfarin, take garlic only on the advice of a herbal or medical practitioner. See also pp.42–51.

Circulatory problems Garlic is today most valued for its positive effects on the circulation. Taken long-term, garlic helps to prevent atherosclerosis (narrowing of the arteries), thins the blood, and supports lower cholesterol levels. These effects promote arterial health, reducing blood pressure and the risk of heart problems.

Longevity Given garlic's wide-ranging health benefits, it is no surprise that in Mediterranean countries it has always been taken to keep one healthy to a ripe old age.

CHOPPED CLOVES

GARLIC OIL

To make garlic oil, crush or finely chop 4 large cloves and place them in a small non-aluminium pan. Add 3 tablespoons of organic olive oil. Heat gently until the mixture is just simmering then simmer for 1–2 minutes, stirring constantly. Pour the contents into a glass jar and leave to cool. Strain, bottle, and label. Use within 12 months.

A bulbous perennial originally from central Asia, garlic is now cultivated commercially worldwide for its use in cooking.

ALOE VERA

Aloe vera

Grown throughout the tropics – and on window sills in cold areas – the thick, spiky, and fleshy leaves of aloe vera yield a cooling gel that makes an excellent treatment for minor burns and abrasions. Known in the United States as a "first aid plant", it should be grown in every kitchen as a handy first aid remedy.

MEDICINAL USES

Part used Leaf

Key actions Anti-inflammatory
● Immune support ● Skin toner
● Wound and tissue healer

Skin conditions, wounds, and burns
Aloe vera has been prized as a medicine for several thousand years, and is now one of the most commonly used of all herbal remedies. Aloe vera's combination of potent healing and anti-inflammatory activity makes it ideal for stimulating repair of damaged tissue, whether resulting from trauma – for example, burns or bruising from a blow – or otherwise. The gel or lotion can be applied neat to sore and inflamed skin, and may be used topically in a wide range of conditions, including acne, dermatitis, herpes sores, nappy rash, nettle rash, psoriasis, radiation burn (after radiotherapy, for example), shingles, and sunburn. For minor burns, first run the affected area under cold tap water for about 10 minutes, then bathe in aloe vera gel. As a rule, aloe vera is best not applied to open wounds. Like the other powerful herbal wound healer comfrey (*Symphytum officinale*), aloe vera has a

gel is a first aid remedy for burns

ALOE VERA GEL

reputation for promoting effective wound healing that minimizes the likelihood of scar (keloid) formation. It combines well with comfrey in healing deep-seated problems such as fractures and sports injuries.

Mouth and throat problems Aloe vera makes an effective wash for all manner of problems occurring within the mouth. Dab the gel onto mouth ulcers or aching teeth. Rinse the mouth and gums daily with gel to help heal gingivitis and to tone receding gums. Aloe vera gel, combined perhaps with

contains healing clear gel

LEAF

KEY INFORMATION

SAFETY	★ ★ ★ ★ ✬
TRADITIONAL USE	★ ★ ★ ★ ★
RESEARCH	★ ★ ★ ★ ☆

BEST TAKEN AS Fresh gel or lotion (topically), prepared juice (internally) ✓✓✓
DOSAGE Aloe vera concentrate: M (*see pp.44–45*)
CAUTIONS The gel can cause an allergic reaction; when using for the first time, apply a small quantity to the skin to test the response. Take internally only on the advice of a herbal or medical practitioner. See also pp.42–51.

Grow aloe vera in a warm location, water sparingly, and you will have an excellent first aid remedy at hand whenever needed. To release the gel, cut the leaf with a sharp knife about 7.5cm (3in) from the tip. On a work surface, carefully slice up the middle of the leaf. Peel back the two sides and expose the clear gel inside. Collect the clear gel and apply as required. Do not use the yellow sap released at the side of the leaf.

FRESH LEAF

sage (*Salvia officinalis*) tea, will make a useful gargle for sore throat and hoarseness, especially where recovery is proving slow.

Other uses Taken internally, aloe vera has long-standing traditional use in the treatment of stomach and duodenal ulcers and irritable bowel syndrome. A growing body of research indicates that aloe vera has a positive stimulant effect on the immune system, with clinical trials suggesting possible benefits in conditions as varied as asthma and HIV. Many species of aloe vera are used in medicine, and some of them are potentially toxic. Quality control is therefore very important when considering taking aloe vera gel internally. For this reason, it is recommended to take aloe vera internally on professional advice only.

Aloe vera thrives in a warm site with indirect sunlight. Like all succulents, it hates being overwatered; allow the soil to dry out in between waterings.

With a long history of use in skin treatments –
Cleopatra attributed her beauty to it – aloe vera
(*Aloe vera*) is today grown worldwide for the
healing clear gel from its leaves.

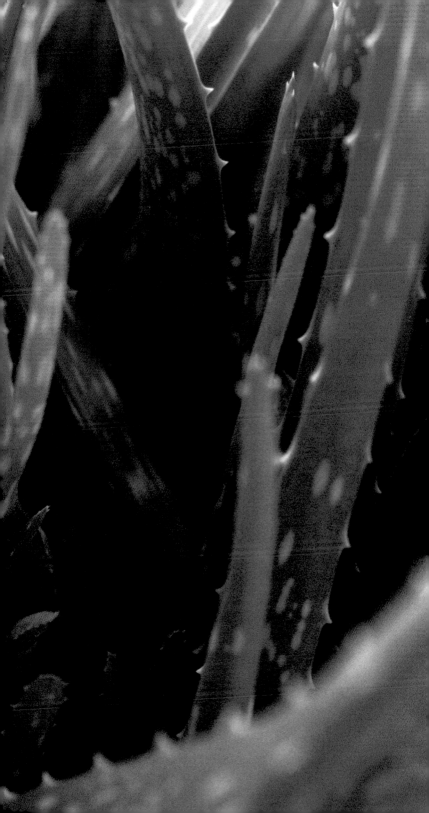

MARSHMALLOW

Althaea officinalis

A traditional European herb, marshmallow has soothing and calming properties that are most used for treating digestive and respiratory disorders.

MEDICINAL USES

Parts used Root • Leaf • Flower

Key actions Demulcent • Emollient • Expectorant

Inflamed mucous membranes
Marshmallow root is typically used to soothe and protect irritated mucous membranes in, for example, acid indigestion, irritable bowel syndrome, and chronic bronchitis. Its sticky consistency means that it works like the body's own mucus to reduce discomfort and inflammation.

Other uses The leaf is preferred for urinary tract problems such as mild cystitis. The flower soothes the skin and contains high levels of antioxidants.

FLOWER

KEY INFORMATION

SAFETY	★ ★ ★ ★ ★
TRADITIONAL USE	★ ★ ★ ★ ☆
RESEARCH	★ ★ ★ ☆ ☆
BEST TAKEN AS	Infusion, decoction ✓✓✓
DOSAGE	A *(see pp.44–45)*
CAUTIONS	None known. See also pp.42–51.

ARNICA

Arnica montana

Widely used in both herbal and homeopathic medicine, arnica's vivid yellow flowers make an excellent remedy for aches and pains of all kinds.

MEDICINAL USES

Part used Flower

Key actions Analgesic (relieves pain) • Anti-inflammatory • Wound healer

FLOWER

Injuries Quick and effective in easing bruises, sprains, and sports injuries, arnica's pain-relieving properties also make it valuable in healing wounds following an operation or dental treatment.

KEY INFORMATION

SAFETY	★ ★ ★ ★ ☆
TRADITIONAL USE	★ ★ ★ ★ ☆
RESEARCH	★ ★ ☆ ☆ ☆
BEST TAKEN AS	Lotion, cream, ointment ✓✓✓
DOSAGE	Topical
CAUTIONS	External use only; toxic when taken internally. Do not apply to broken skin or open wounds. See also pp.42–51.

ANGELICA

Angelica archangelica

The botanical name of angelica points to the highly prized status of this northern European herb in the past. A warming tonic that is good alike for poor digestion and weak circulation, angelica is also an excellent remedy to support recovery from chronic illness and to revitalize a delicate digestive system.

MEDICINAL USES

Parts used Root ● Seed

Key actions Expectorant ● Muscle relaxant (mild) ● Relieves wind ● Stimulates appetite and digestive juices ● Stimulates sweating and cooling ● Strengthens weak circulation

Digestive system The bitter taste of angelica – best savoured as a tincture – stimulates stomach activity, making it a key remedy for poor appetite and anorexia. It soothes cramping and sensations of fullness in the digestive tract, and eases wind.

DRIED ROOT

Respiratory problems Angelica serves well in conditions such as asthma, bronchitis, chest congestion, and cough, and is an ideal remedy for recuperation after an acute chest infection.

KEY INFORMATION

SAFETY   ★ ★ ★ ★ ☆
TRADITIONAL USE   ★ ★ ★ ★ ☆
RESEARCH   ★ ☆ ☆ ☆ ☆
BEST TAKEN AS Tincture ✓✓✓ Capsule ✓✓ Tablet ✓
DOSAGE C (*see pp.44–45*)
OFTEN USED WITH Chamomile (*Chamomilla recutita*)
CAUTIONS Do not take in pregnancy or if taking anticoagulants (blood-thinning drugs). Not advisable during heavy menstrual bleeding. See also pp.42–51.

Other uses The root's warming and stimulant action upon the heart and circulation improves blood flow throughout the whole body, making it useful in problems such as cold hands and feet, chilblains, and fibromyalgia.

Found in temperate regions as far apart as western Europe, Siberia, and the Himalayas, angelica grows in damp sites.

CHINESE ANGELICA, DONG QUAI

Angelica sinensis

One of the most popular of all Chinese herbs, Chinese angelica is regarded as the main women's tonic, helping to support a regular menstrual cycle and easing menstrual pain. As a warming and relaxing remedy that strengthens digestive and liver activity, it is suitable for both men and women.

MEDICINAL USES

Part used Root

Key actions Digestive tonic ● Female reproductive tonic ● Relieves menstrual pain

root is used as a uterine tonic

DRIED ROOT

Menstrual problems Also known as dong quai, Chinese angelica may be taken to help maintain a normal menstrual cycle or to treat menstrually linked problems such as breast tenderness and painful periods. In the case of irregular or missed periods, it will often help to establish a more regular menstrual cycle if taken for several months. However, it should be avoided where heavy menstrual bleeding occurs. Although Chinese angelica appears to have no direct hormonal activity, it has the reputation of helping to improve fertility, combining well with chaste berry (*Vitex agnus-castus*) in this respect.

TINCTURE

Other uses Preliminary research suggests that Chinese angelica could be useful in helping to prevent or slow osteoporosis.

KEY INFORMATION

SAFETY	★ ★ ★ ★ ☆
TRADITIONAL USE	★ ★ ★ ★ ★
RESEARCH	★ ★ ☆ ☆ ☆

BEST TAKEN AS Decoction ✓✓✓
Tincture ✓✓ Capsule, tablet ✓
DOSAGE C *(see pp.44–45)*
OFTEN USED WITH Chaste berry
(*Vitex agnus-castus*)
CAUTIONS Do not take during pregnancy or if taking anticoagulant (blood-thinning) medication. Not advisable during heavy menstrual bleeding. See also pp.42–51.

Pleasant-tasting and with a slightly peppery flavour, Chinese angelica is a common ingredient in Chinese medicinal food dishes.

CELERY

Apium graveolens

A good detox remedy, celery stem, leaf, and seed stimulate the kidneys to clear waste products, especially helping to cleanse salts that accumulate in joints, causing stiffness and inflammation. Although celery today is considered to be a mild sedative, in earlier times it was believed to be an aphrodisiac.

MEDICINAL USES

Parts used Leaf ● Stem ● Seed

Key actions Anti-inflammatory ● Anti-rheumatic ● Relieves wind ● Stimulates urine flow

Arthritic and rheumatic problems

Celery is a key remedy in European herbal medicine in the treatment of arthritic and rheumatic problems where joints, muscles, and tendons are sore, swollen, or stiff. The seeds contain a volatile oil which stimulates the clearance of waste products by the kidneys. In particular, celery supports the elimination of salts such as urates that often cause inflammation and stiffness within the muscular-skeletal system. Celery seeds are taken to ease joint and muscle pain and stiffness, especially when it occurs in the early

morning, as well as to clear fluid accumulation linked to arthritis. Gout is one of its main indications – the seeds can be taken to relieve symptoms and prevent recurrence.

Detox remedy Juiced or in a smoothie, the stem and leaf are key ingredients in any regimen looking to support detoxification, clear fluid retention, or aid weight loss.

Other uses Celery seeds may also be taken to relieve wind and bloating.

KEY INFORMATION	
SAFETY	★★★★☆
TRADITIONAL USE	★★★★☆
RESEARCH	★★☆☆☆
BEST TAKEN AS	Tincture ✓✓✓
	Capsule ✓✓ Tablet ✓
DOSAGE	D (*see pp.44–45*)

OFTEN USED WITH Willow bark (*Salix alba*)
CAUTIONS Seeds not to be taken during pregnancy or in kidney disease. Occasionally, can cause allergic reactions, including contact dermatitis. See also pp.42–51.

a nutritious vegetable — **STEM AND LEAVES**

seeds contain volatile oil —

DRIED SEEDS

Both the juice and seeds of celery stimulate urine flow and can prove useful as part of a broad approach to treating high blood pressure.

One of the foremost detoxifying herbs in Western and Chinese herbal medicine, burdock (*Arctium lappa*) is used to cleanse the body of waste products, including heavy metals.

BURDOCK

Arctium lappa

Traditionally combined with dandelion to make a tonic, cleansing drink, burdock is an important detox remedy in both western and Chinese herbal traditions. Often used to treat skin problems, burdock also supports the immune system during infection and chronic illness.

MEDICINAL USES

Parts used Root ● Leaf ● Seed

Key actions Antiseptic ● Detoxifying remedy ● Diuretic ● Tonic

Blood cleanser Although much undervalued, burdock may be used in any situation in which the body needs increased clearance of waste products. Conditions as varied as acne, boils, eczema, arthritis, fibromyalgia, and tonsillitis will benefit from the herb's pronounced ability to stimulate release of waste products from the cells.

seeds have cleansing and diuretic properties

SEEDS   DRIED ROOT

However, it should be used with caution as even small amounts can cause an initial flare-up in symptoms, particularly in skin disorders. For this reason, it is rarely used on its own and is combined with remedies such as dandelion (*Taraxacum officinale*), red clover (*Trifolium pratense*), and yellow dock (*Rumex crispus*) that counterbalance its detoxifying action. Burdock root is also thought to have anti-diabetic and cancer-preventative activity.

KEY INFORMATION

SAFETY	★ ★ ★ ★ ☆
TRADITIONAL USE	★ ★ ★ ★ ☆
RESEARCH	★ ☆ ☆ ☆ ☆

BEST TAKEN AS Decoction ✓✓✓
Tincture ✓✓ Tablet ✓
DOSAGE A (*see pp.44–45*)
OFTEN USED WITH Yellow dock
(*Rumex crispus*)
CAUTIONS Very rarely, can cause contact dermatitis. See also pp.41–52.

Leaves can be used as a poultice for acne and boils.

ASTRAGALUS

Astragalus membranaceus

This remarkable herb has been used in Chinese herbal medicine for over 2,000 years, and scientific research is beginning to confirm (and to some degree, extend) its range of uses. Astragalus is a safe remedy, often helpful in cases of chronic infection.

MEDICINAL USES

root supports immune system

DRIED ROOT

Part used Root

Key actions Antioxidant ● Immune support ● Tonic

Low endurance and weak immune resistance Seen as a specific remedy for supporting a weak or compromised immune system, astragalus is also classified as an adaptogen, strengthening the body's ability to cope with the physical aspects of stress. The root is used to treat many longstanding health problems, especially those that involve chronic infection, weakness, and exhaustion. Chronic fatigue syndrome, viral infections, and debility can all benefit from medium to long-term use of the herb. Astragalus

KEY INFORMATION

SAFETY ★★★☆☆
TRADITIONAL USE ★★★★★
RESEARCH ★★★★☆
BEST TAKEN AS Decoction ✓✓✓
Tincture ✓✓ Capsule, tablet ✓
DOSAGE A (*see pp.44–45*)
OFTEN USED WITH Schisandra (*Schisandra chinensis*).
CAUTIONS No known side effects. Avoid in acute illness. See also pp.42–51.

also has a persistent reputation for helping to control sweating, especially where this is linked to chronic illness.

Cancer Although astragalus is not a treatment for cancer in its own right, it nevertheless has much to offer in supporting immune function and maintaining a sense of well-being. It is best taken either as a preventative or alongside conventional treatment such as chemotherapy. Clinical research suggests that astragalus reduces the toxic effects of chemotherapy and at the same time enhances immune function. The root may need to be taken long-term to achieve best results. Seek professional advice from your doctor or herbal practitioner before taking astragalus along with chemotherapy.

Native to China and Mongolia, astragalus is one of the most popular tonic herbs in China and is used to improve energy levels and increase endurance.

OATS

Avena sativa

Better known as a food, oats are a valuable medicinal plant with health-giving effects on the nervous system. A good source of both B vitamins and vitamin E, oats are absorbed slowly into the blood stream, have a low glycaemic index, and support better-balanced blood sugar levels.

MEDICINAL USES

Parts used Dried seed • Fresh plant

Key actions Antidepressant • Emollient • Nutritive • Tonic

Nervous problems Oats can be taken therapeutically to improve nervous stamina and lift depressed mood. Traditional use ascribes antidepressant activity to the dried seeds and fresh plant, and either may be useful where lowered mood is associated with anxiety and nervous exhaustion, especially during the menopause. The fresh plant is a tonic remedy for all types of nervous debility, and can help to improve sleep duration and quality where the person is literally too tired to sleep. Oats also aid withdrawal from tobacco and drug addiction.

seeds have a mild antidepressant activity

DRIED SEEDS (GRAIN)   **FRESH STEM**

Eczema The dried seeds can be used to make a decoction to relieve the symptoms of eczema. Put the seeds in a muslin bag under a running hot bath tap so that the decoction is strained into the bath – the soothing emollient activity of the seeds eases itching and nourishes the skin.

Oats are grown principally as a cereal crop. Their tonic action on the nervous system has led herbalists to describe them as food for the nerves.

KEY INFORMATION

SAFETY	★ ★ ★ ★ ★
TRADITIONAL USE	★ ★ ★ ✬ ☆
RESEARCH	★ ★ ✬ ☆ ☆

BEST TAKEN AS Dried seed: Food ✓✓✓ Tincture ✓✓ Fresh plant: Tincture ✓✓✓ Capsule ✓✓

DOSAGE A (*see pp.44–45*)

OFTEN USED WITH Damiana (*Turnera diffusa*) for nervous exhaustion or stress with depressed mood.

CAUTIONS None known. See also pp.42–51.

NEEM

Azadirachta indica

A large evergreen tree, neem is a veritable pharmacy in its own right, as well as being a natural insecticide. The seed, seed oil, leaf, and bark are used medicinally, and have found use in conditions as diverse as scabies and psoriasis, malaria, diabetes, and anxiety.

MEDICINAL USES

Parts used Bark ● Leaf ● Seed ● Twig

Key actions Antibacterial ● Antifungal ● Anti-inflammatory ● Blood cleanser ● Immune support ● Lowers blood sugar levels ● Relieves itchiness

Skin conditions Neem's most common use in the West is as an oil, which can bring relief to sore and itchy rashes. It may be safely applied to irritated or inflamed skin such as in eczema and psoriasis, and can be used to treat headlice, scabies, and fungal problems such as ringworm. The oil may also be applied as a poultice to boils, helping to draw out toxins.

Other uses Taken internally, neem's many uses include bacterial, fungal, and viral infection, allergic reactions such as asthma, diabetes, digestive problems such as peptic ulcer, and liver disorders. A strong-acting remedy, neem appears to be safe at normal dosage, but it is best taken internally when prescribed by a herbal practitioner or doctor. Neem has contraceptive activity and should be avoided by women who wish to conceive.

Planted in villages and towns throughout India, neem often acts as a community medicinal resource and is one of the most valued herbs in Ayurvedic medicine.

a leaf infusion is used for skin rashes

FRESH LEAVES

KEY INFORMATION

SAFETY   ★ ★ ☆ ☆ ☆
TRADITIONAL USE   ★ ★ ★ ★ ★
RESEARCH   ★ ★ ★ ⯪ ☆
BEST TAKEN AS Tincture ✓✓✓ Capsule ✓✓ Tablet ✓
DOSAGE C (*see pp.44–45*)
CAUTIONS Do not take during pregnancy, while breast-feeding, or during fertility treatment. In children, use topically *only*. Keep to recommended dosage – long-term high-dose use is not advisable. See also pp.42–51.

BACOPA

Bacopa monniera

An important Ayurvedic herb, bacopa is a specific for reducing nervous overactivity and improving mental performance.

MEDICINAL USES

Part used Dried whole plant

Key actions Mild analgesic ● Mild sedative ● Nerve tonic

KEY INFORMATION

SAFETY	★ ★ ★ ★ ☆
TRADITIONAL USE	★ ★ ★ ★ ★
RESEARCH	★ ★ ⯪ ☆ ☆

BEST TAKEN AS Capsule ✓✓✓ Tablet ✓✓ Tincture ✓
DOSAGE B *(see pp.44–45)*
CAUTIONS Can cause digestive irritation. See also pp.42–51.

Nervous and digestive disorders

A gentle-acting remedy for nervous exhaustion, anxiety, and stress, bacopa also benefits digestion, cooling excess heat and stimulating appetite. Traditional usage and scientific research suggest that bacopa can help to enhance memory and concentration. It also appears to improve learning ability.

WHOLE PLANT

BUCHU

Barosma betulina

A key tonic, antiseptic, and mild stimulant herb in South African traditional medicine, buchu helps to relieve urinary tract infections and windy digestions.

MEDICINAL USES

Part used Leaf

Key action Urinary antiseptic

Urinary tract infection Buchu is a specific for cystitis and infection within the urinary tract as a whole, its essential oil having marked antiseptic activity. The herb is best taken as an infusion and is probably most effective when used short term for acute infections. Other herbs such as cranberry (*Vaccinium macrocarpon*) may be better for chronic conditions.

KEY INFORMATION

SAFETY	★ ★ ★ ★ ☆
TRADITIONAL USE	★ ★ ★ ★ ⯪
RESEARCH	★ ☆ ☆ ☆ ☆

BEST TAKEN AS Infusion ✓✓✓ Tincture ✓✓ Capsule ✓
DOSAGE B *(see pp.44–45)*
OFTEN USED WITH Corn silk (*Zea mays*)
CAUTIONS During pregnancy, take only on the recommendation of a herbal or medical practitioner. Potentially toxic at excess dosage. See also pp.42–51.

leaves contain antiseptic volatile oil

DRIED LEAVES

OREGON GRAPE

Berberis aquifolium syn. *Mahonia aquifolium*

A strongly bitter-tasting herb from Pacific northwest America, Oregon grape has a long history as a digestive tonic and appetite stimulant. Over the last 20 years evidence has grown supporting its use in treating chronic skin disorders.

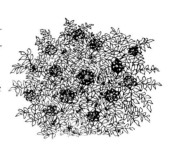

MEDICINAL USES

Part used Root

Key actions Bitter tonic ● Mild laxative ● Skin cleanser

Digestive disorders In common with its close relative, barberry (*Berberis vulgaris*), Oregon grape exerts a directly positive effect on the digestive system – inflammatory problems within the stomach or gall bladder, poor appetite, and indigestion are key indications for the herb.

Skin conditions Oregon grape is most commonly used to treat chronic inflammatory or infected skin conditions such as acne, eczema, and psoriasis. The herb contains constituents known to slow down excessive skin growth and to have antibacterial and antifungal activity.

Abundant in Oregon and northern California, Oregon grape grows at high altitudes in the Rocky Mountains as well as the Pacific coastal regions.

Clinical trials have found that Oregon grape extract, cream, or ointment help to relieve psoriasis. Best results are likely to be obtained by using Oregon grape in combination with other remedies that have established activity in treating chronic skin disorders.

DRIED ROOT

KEY INFORMATION

SAFETY	★ ★ ★ ⯪ ☆
TRADITIONAL USE	★ ★ ★ ☆ ☆
RESEARCH	★ ★ ★ ⯪ ☆

BEST TAKEN AS Tincture ✓✓✓ Decoction ✓✓ Capsule ✓

DOSAGE B (*see pp.44–45*)

OFTEN USED WITH Liquorice (*Glycyrrhiza glabra*)

CAUTIONS Do not take during pregnancy and while breast-feeding. See also pp.42–51.

A Mediterranean plant from Spain and Morocco, borage *(Borago officinalis)* is not only grown as a garden herb and decorative plant but is also extensively cultivated for its seed oil.

BEETROOT, RED BEET

Beta vulgaris

Used as a medicinal food since Roman times, the earthy-tasting beetroot and its juice support liver function and the removal of waste products from the body.

MEDICINAL USES

Part used Root ● Leaf

Key actions Antioxidant ● Nutritive ● Lowers blood pressure

Aids liver health, lowers blood pressure Beetroot promotes liver regeneration, maintains healthy fat metabolism, and helps to stabilize blood sugar levels. Recent research shows that beetroot juice lowers blood pressure in people with high blood pressure.

Red rather than white beetroot is best for supporting the heart and circulation.

KEY INFORMATION

SAFETY   ★ ★ ★ ★ ★
TRADITIONAL USE   ★ ★ ★ ★ ★
RESEARCH   ★ ★ ★ ⯪ ☆
BEST TAKEN AS Vegetable ✓✓✓ Juice ✓✓✓
DOSAGE 1–2 glasses of juice a day
OFTEN USED WITH Celery (*Apium graveolens*)
CAUTIONS None known

BORAGE, STARFLOWER

Borago officinalis

A native European and north African annual, borage seed – with up to 25 per cent oil content – is widely used for its omega-6 fatty acids.

MEDICINAL USES

Part used Seed oil

Key actions Anti-inflammatory ● Antioxidant ● Emollient (soothes skin)

Chronic skin disorders With its high omega-6 fatty acid content, borage oil has significant anti-inflammatory activity, and taken over several months can improve skin conditions such as eczema. Apply locally to dry or itchy skin patches.

KEY INFORMATION

SAFETY   ★ ★ ★ ★ ★
TRADITIONAL USE   ★ ★ ☆ ☆ ☆
RESEARCH   ★ ★ ★ ★ ☆
BEST TAKEN AS Oil ✓✓✓ Capsule ✓✓
DOSAGE M (*see pp.44–45*)
CAUTIONS Seed oil: best taken with food. Seek advice from a herbal or medical practitioner if taking epilepsy medication. Avoid other borage products, which are thought to contain constituents that are potentially toxic to the liver. See also pp.42–51.

FLOWERS

BIRCH

Betula spp.

The birch tree has a long history of use in northern temperate regions of the world. Birch tar oil, distilled from the bark, is a traditional treatment for chronic skin diseases. The leaves are used in kidney and rheumatic disorders, and the sap, tapped in early spring, is taken as a refreshing and cleansing tonic.

MEDICINAL USES

Parts used Bark ● Leaf

Key actions Anti-inflammatory ● Astringent ● Diuretic ● Mild analgesic ● Stimulates sweating (diaphoretic)

Rheumatic and kidney problems

Birch's unusual combination of actions makes it a valuable remedy in conditions where symptoms reflecting kidney weakness – poor urine output, fluid retention, and puffiness – occur side by side with rheumatic problems such as stiff and aching muscles, arthritic pain, and leg cramps. By aiding the clearance of waste products in urine, birch leaves increase the body's ability to remove waste products from joint and muscle tissues. The leaves contain aspirin-like substances that contribute to their ability to control inflammation and relieve pain. Traditional uses of birch include rheumatic pain, gout, fibromyalgia, and kidney and urinary tract infections such as cystitis.

A warm decoction of the leaves and twigs can be applied to ease stiff and aching muscles. The sap is thought to have diuretic properties.

KEY INFORMATION

SAFETY	★ ★ ★ ★ ☆
TRADITIONAL USE	★ ★ ★ ☆ ☆
RESEARCH	★ ★ ☆ ☆ ☆

BEST TAKEN AS Infusion ✓✓✓ Tincture ✓✓ Capsule ✓
DOSAGE A (*see pp.44–45*)
OFTEN USED WITH Willow bark (*Salix alba*)
CAUTIONS None known.
See also pp.41–52.

Other uses

A favourite Scandinavian remedy, birch twig bundles are used in saunas and steam baths to beat the skin and muscles in order to stimulate sweating, invigorate, and relieve tender and aching muscles. Birch oil, extracted from the leaves and twigs, is a traditional northern European product, commonly used in external applications for rheumatic aches and pains.

DRIED LEAVES

Few trees have proved to be more useful than the birch: in addition to its medicinal qualities, the bark is used as tinder and paper, and the trunk is made into buckets and canoes.

BOSWELLIA

Boswellia serrata

Boswellia is highly prized in traditional Indian medicine and has been used to treat conditions as varied as arthritis, asthma, bronchitis, dysentery, and fever. The rationale for its traditional use has been largely confirmed by recent scientific research.

MEDICINAL USES

Part used Resin

Key actions Anti-inflammatory
● Anti-arthritic ● Antiseptic
● Reduces fever

Arthritic problems Boswellia is fast becoming one of the most commonly taken medicines for arthritic problems. Concerns over the safety of conventional

KEY INFORMATION

SAFETY	★ ★ ★ ★ ☆
TRADITIONAL USE	★ ★ ★ ★ ☆
RESEARCH	★ ★ ★ ★ ☆

BEST TAKEN AS Tablet: standardized to 60% boswellic acid content ✓✓✓
DOSAGE M *(see pp.44–45)*
OFTEN USED WITH Ginger (*Zingiber officinale*)
CAUTIONS Can cause contact dermatitis. See also pp.42–51.

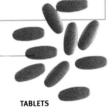

standardized extract in tablet form

TABLETS

anti-inflammatories have increased interest in herbal alternatives, and in boswellia's case, there is a significant and growing body of research that indicates both its safety and effectiveness. The specific anti-inflammatory action of the resin makes it an important remedy for chronic inflammatory conditions such as rheumatoid arthritis and gout. It can also prove valuable in relieving pain and stiffness in osteoarthritis.

Other uses Boswellia is also indicated in other inflammatory conditions such as asthma, ulcerative colitis, and multiple sclerosis. It has recently been used to treat brain tumours and Alzheimer's disease – in both cases, it should be used only under professional supervision.

Boswellia is a relative of frankincense (*Boswellia sacra*), and is sometimes known as Indian frankincense. It grows in India and North Africa.

BUPLEURUM, CHAI HU

Bupleurum falcatum syn. *B chinensis*

A member of the carrot family, bupleurum root is a commonly used remedy in China and Japan, its main traditional applications being flu, flu colds and fever, irregular menstrual cycle, and liver disorders. It is a good bitter-tasting tonic and, as with all bitter remedies, stimulates appetite and digestive function.

MEDICINAL USES

root supports liver function

DRIED ROOT

Part used Root

Key actions Anti-inflammatory ● Bitter tonic ● Protects liver ● Stimulates sweating

Immune function Ongoing research over the last 30 years – mostly in Japan – suggests that bupleurum has a unique combination of medicinal benefits. It has potent anti-inflammatory activity, similar in some ways to steroid medication, which helps reduce and prevent inflammation throughout the body. At the same time, bupleurum enhances immune function and protects both liver and kidneys from damage. Its key use is therefore in liver and kidney disorders, especially where these are under stress due to chronic inflammation, toxicity, or autoimmune disease. It may be safely self-medicated for straightforward health problems, for example in treating colds, flu, and fever, but should be taken only on professional recommendation in serious illness such as autoimmune disease.

KEY INFORMATION

SAFETY ★ ★ ★ ★ ☆
TRADITIONAL USE ★ ★ ★ ★ ★
RESEARCH ★ ★ ☆ ☆
BEST TAKEN AS Decoction ✓✓✓
Tincture ✓✓ Tablet ✓
DOSAGE B (*see pp.44–45*)
OFTEN USED WITH Astragalus (*Astragalus membranaceus*)
CAUTIONS Can occasionally cause gastro-intestinal symptoms such as flatulence, nausea, and vomiting.
See also pp.42–51.

DECOCTION

Other uses In traditional Chinese medicine, bupleurum is combined with herbs such as white peony (*Paeonia lactiflora*) and liquorice (*Glycyrrhiza glabra*) to treat conditions such as irregular menstrual activity and prolapsed womb.

Bupleurum's use in China as a liver tonic extends back at least 2,000 years. The root is dug up in spring or autumn when it contains the most active constituents.

Difficult to mistake for any other plant, calendula (*Calendula officinalis*) is instantly recognized by its bright orange flowerheads. The flowers are used to heal rashes and inflamed, sore skin.

MARIGOLD, CALENDULA

Calendula officinalis

Calendula is best known as a cream or ointment that makes a soothing and healing application to sore, angry, or inflamed skin. The herb's bright orange flowerheads can also be prepared as an infusion that, once cooled, makes a soothing wash or lotion for hot and inflamed rashes, cuts, or grazes.

MEDICINAL USES

Part used Flower

Key actions Antifungal
- Anti-inflammatory • Antimicrobial
- Blood cleanser • Wound healer

Skin infections Whether applied topically on the skin or taken internally, calendula has antiseptic, cleansing, and detoxifying properties, and a wealth of potential uses. As a lotion, cream, or ointment, it will speed healing and counter infection in conditions as diverse as minor burns and sunburn, insect bites and stings, sore and pustular spots, mastitis, cuts and abrasions, inflamed rashes, haemorrhoids, and varicose veins.

Digestive disorders Taken internally (best as an infusion), calendula may be used to help to heal inflammatory problems throughout the digestive tract, including peptic ulcer

AERIAL PARTS

Cultivated in temperate regions, calendula grows well in almost all soils. The flowers are harvested in early summer when they open.

and gastritis. Due in part to its antifungal properties, it will aid recovery from gastrointestinal infection, especially when linked to gut dysbiosis and candidiasis, and help to treat problems such as acne, throat infections, and mastitis. As an infusion, it combines well with herbs such as cleavers *(Galium aparine)*, red clover *(Trifolium pratense)*, and chamomile *(Chamomilla recutita)*.

KEY INFORMATION

SAFETY	★★★★☆
TRADITIONAL USE	★★★★☆
RESEARCH	★★½☆☆

BEST TAKEN AS Infusion ✓✓✓
Tincture ✓✓ Capsule ✓
DOSAGE B *(see pp.44–45)*
OFTEN USED WITH Cleavers *(Galium aparine)*
CAUTIONS Can occasionally cause allergic reactions. See also pp.42–51.

flowers have antiseptic properties

DRIED FLOWERS

SENNA

Cassia spp.

A well-known herb, senna grows in much of North Africa, the Middle East, and India, and its use has become almost universal. It was first used by Arab physicians in the 9th century CE. With its strong laxative action, senna makes an effective short-term treatment for constipation.

MEDICINAL USES

Parts used Leaf ● Pod

Key actions Stimulant laxative

Constipation Senna is primarily used to treat acute and short-term constipation – and it usually achieves this end efficiently. Senna is best taken in the evening, as active constituents within both leaf and pod irritate the muscles of the colon and normally result in a bowel movement 6–8 hours later. The standard advice is to take senna for up to a maximum of two weeks at a time. If constipation remains a problem after two weeks, seek advice from your herbal practitioner or doctor. To minimize the chances of griping, senna should be combined with a relaxant remedy such as chamomile (*Chamomilla recutita*), fennel (*Foeniculum vulgare*), or ginger (*Zingiber officinale*). At the appropriate dosage, senna is a very safe medicine. Not only is it safe to take during pregnancy and while breast-feeding, but it seems to be the laxative of choice to relieve the constipation that commonly occurs during pregnancy.

KEY INFORMATION

SAFETY	★ ★ ★ ★ ✰
TRADITIONAL USE	★ ★ ★ ★ ★
RESEARCH	★ ★ ★ ★ ★

BEST TAKEN AS Tablet ✓✓✓ Syrup ✓✓
DOSAGE M (*see pp.44–45*)
OFTEN USED WITH Ginger (*Zingiber officinale*)
CAUTIONS Take at recommended dosage only – may cause abdominal cramping. See also pp.42–51.

Native to Africa, senna is a small shrub with a woody stem.

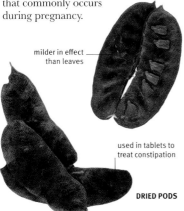

milder in effect than leaves

used in tablets to treat constipation

DRIED PODS

TEA

Camellia sinensis

Grown almost exclusively for use as a beverage, tea is perhaps the world's most undervalued medicinal plant. Numerous studies point to the health-giving properties of the tea leaf, especially unfermented green or white tea. These teas contain high levels of polyphenols, which have a potent antioxidant activity.

MEDICINAL USES

Parts used Leaf bud ● Young leaf

Key actions Antioxidant ● Astringent ● Diuretic ● Stimulant

Stimulant Traditionally, tea has been seen as a gentle stimulant, its moderate caffeine content enhancing mental alertness and acting as a "pick-me-up". Like coffee (*Coffea arabica*), it has been used as a remedy for headache, though coffee is probably more effective in this respect. Tea has warming and tonic properties, endearing it to those working in the cold. It is best avoided in premenstrual syndrome – research has repeatedly shown that caffeine leads to a worsening of symptoms, and may be unhelpful during the menopause, since it can increase hot flushes.

Digestive problems As an astringent, tea makes a useful and readily available remedy for diarrhoea; the polyphenols in the tea counter infection and tone up the inner lining of the gut.

green tea prevents tooth decay

FRESH LEAVES

Recent studies suggest that the polyphenolic compounds in tea may promote fertility in women who drink one to two cups daily.

In combination with other remedies, traditional Chinese medicine uses tea to treat diarrhoea and dysentery.

Eye problems For tired, irritated, and puffy eyes, place a moist tea bag or cotton wool that has been soaked in cooled green tea on the affected (closed) eye for a few minutes. Usually symptoms are eased and the eyelids and surrounding tissues are toned. Tea can also be used in this way to counter inflammation or

brew of tea leaves soothes sunburn

DRIED LEAVES

KEY INFORMATION

SAFETY	★ ★ ★ ★ ★
TRADITIONAL USE	★ ★ ★ ★ ☆
RESEARCH	★ ★ ★ ★ ☆

BEST TAKEN AS Infusion ✓✓✓ Tablet ✓✓ Capsule ✓

DOSAGE B (see pp.44–45)

CAUTIONS Avoid excessive doses. See also pp.42–51.

infection within the eye, for example in helping to relieve the pain and discomfort of conjunctivitis.

Other uses Recent research has focused on the antioxidant polyphenols, which have been found to aid weight loss, to counter inflammation, and to have anti-cancer and anti-tumour activity. The high intake of green tea in China and Japan is thought to be partly responsible for the low incidence of cancer in these countries. Tea also appears to reduce the incidence of tooth decay. Not surprisingly, green tea has become a popular drink in the West in recent years, though it is worth noting that research suggests that tea impairs absorption of iron and other minerals. Avoid drinking tea with meals or medication, especially if anaemic.

AGE-OLD BREW

One of the legends surrounding tea drinking features Shen Nung, a 3rd-century BCE Chinese emperor. It is said that as he sat under a tree, boiling his drinking water, a few leaves from the tree (*Camellia sinensis*) fell into his kettle. Shen Nung drank the brew and found the taste quite agreeable, thus prompting the long-standing tradition of tea drinking.

TEA CADDY

Tea quality Green tea is produced by lightly steaming the freshly cut bud and leaf, leaving the active constituents largely intact. However, black tea is allowed to ferment, leading to a significant loss of antioxidant constituents, notably polyphenols. High-quality teas – for example white tea, which is made with the youngest leaf buds – have the highest levels of polyphenols and can be expensive, though not as costly as in the 18th century, when the finest teas were literally worth their weight in gold.

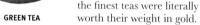

Tea is cultivated mainly in India, Sri Lanka, and China. Tea leaves are picked throughout the year.

GREEN TEA

CHILLI, CAYENNE PEPPER

Capsicum spp.

Familiar the world over, chilli comes from the Americas, its strongly pungent flavour giving spice to countless dishes. The constituents responsible for the hot, sometimes fiercely hot, impact of chilli when eaten are also those most involved in its many medicinal applications.

MEDICINAL USES

Part used Fruit

Key actions Antiseptic ● Counter-irritant ● Local analgesic ● Relieves wind and spasm ● Stimulant ● Tonic

Circulation When it is applied on the skin as a counter-irritant, chilli, like other hot, pungent remedies such as mustard (*Sinapis alba*), causes irritation and swelling, and an increase in circulation to the area. Chilli is added to lotions, liniments, and salves for muscular aches and pains for this reason, resulting in better nutrition to – and clearance of waste products from – the tissue involved.

Introduced to Europe in the 16th century, chilli is now cultivated throughout the tropics, particularly in Africa and India.

Nerve pain Chilli peppers are graded according to their "heat"; the hotter the taste, the higher the level of capsaicin – the key active constituent present within the flesh of the pepper. Capsaicin initially increases the awareness of pain and inflammation, but then desensitizes the local nerve endings, leading to reduced levels of pain. This action is utilized in capsaicin creams for conditions such as post-herpetic neuralgia (shingles), nerve pain linked to diabetes, and also for severe itchiness. These products are often only available on prescription.

General stimulant Rarely used on its own, chilli is most commonly added to other herbs to strengthen and stimulate their action within the body. Adding small quantities of chilli powder, sauce, or tincture can provide an important

boost to the effects of a herbal mixture. For example, a few drops of chilli sauce or tincture can be mixed with echinacea (*Echinacea* spp.) and liquorice (*Glycyrrhiza glabra*) tincture to treat throat infection. To strengthen the circulation and to improve blood flow to hands and feet, small quantities of chilli can be routinely added to food. Chilli also mixes well with specific remedies such as devil's claw (*Harpagophytum procumbens*) when treating conditions such as osteoarthritis and fibromyalgia, where circulation to affected areas is often poor. Chilli's general stimulant effect also finds use where the thyroid gland is mildly underactive. Here, chilli will help to strengthen the circulation and improve metabolic rate. Recent research points to the possibility that chilli works to enhance the anti-cancer activity of other antioxidant remedies. Although data is based so far only on test tube research, scientists found that when they combined 1 part chilli preparation with 25 parts green tea concentrate, the anti-cancer activity of the combination was 100 times greater than the green tea alone. This suggests that chilli can have a major impact on the medicinal activity of other foods and herbal remedies.

KEY INFORMATION

SAFETY	★ ★ ★ ★ ☆
TRADITIONAL USE	★ ★ ★ ★ ★
RESEARCH	★ ★ ★ ☆ ☆

BEST TAKEN AS Capsule ✓✓✓
DOSAGE C (*see pp.44–45*)
CAUTIONS Non-toxic at normal dose; caution required when eating or handling hot chilli products. Can cause intense pain and burning, and contact dermatitis. See also pp.42–51.

CHILLI POWDER

Weak digestive system Chilli has antiseptic properties and helps to protect against gastrointestinal infection. It is often added to food in tropical countries to reduce the risk of food poisoning. Used in small quantities, chilli will help to strengthen a weak digestive system and stimulate appetite, particularly in older people, though it can benefit anyone with an under-performing digestive system. A pinch of chilli powder added to chamomile (*Chamomilla recutita*) tea or tincture will help to relieve nausea and feelings of fullness. Surprisingly, chilli can be an effective remedy for treating diarrhoea.

CHILLI "PEPPER"

Chilli was one of many medicinal plants Christopher Columbus brought back from the New World to present to his patrons, the Spanish king and queen. The hot taste of chilli, similar to that of black pepper from Indonesia, encouraged Columbus to call it chilli "pepper", thereby suggesting that it also came from eastern Asia, and that his mission to find a western route to the East Indies had been successful.

fruit is used to promote local circulation

FRESH FRUIT

tincture can boost action of other herbal mixtures

TINCTURE

NEW WORLD SPICE

The hot taste of chilli (*Capsicum* spp.) is a pointer to its medicinal use as a powerful warming stimulant. A popular ingredient in Mexican cuisine, where it is even used to flavour ice-cream.

PAPAYA, PAW PAW

Carica papaya

A sweet-tasting fruit and native to tropical America, papaya while unripe contains digestive enzymes that complement the body's own digestive juices.

MEDICINAL USES

Parts used Fruit pulp ● Latex (extract)

Key actions Digestive ● Nutritive

Digestive problems When the unripe fruit is cut, a thick white juice or latex containing papain (digestive enzyme) seeps out. Papain breaks down protein, especially in an alkaline environment, making it a useful supplement that promotes effective digestion within the small intestine, in particular when normal digestive secretions are deficient.

RIPE FRUIT

KEY INFORMATION

SAFETY	★ ★ ★ ★ ★
TRADITIONAL USE	★ ★ ★ ✬ ☆
RESEARCH	★ ★ ★ ☆ ☆

BEST TAKEN AS Extract: Tablet, capsule ✓✓✓
DOSAGE M – extract *(see pp.44–45)*
CAUTIONS Caution required in pregnancy when using concentrated forms of the papaya enzyme, papain. See also pp.42–51.

Papain is also used as a food tenderizer, especially in the fast-food industry.

Other uses The ripe fruit is nutritious, cleansing, and mildly laxative. Papaya seed can be used to treat worms.

CARAWAY

Carum carvi

Used in food and medicine for at least 5,000 years, caraway is one of Europe's most popular herbs. The volatile oil in the seeds gives caraway its distinctive aroma.

MEDICINAL USES

Parts used Seed ● Essential oil

Key actions Eases coughing ● Relieves wind and spasm

KEY INFORMATION

SAFETY	★ ★ ★ ★ ☆
TRADITIONAL USE	★ ★ ★ ★ ☆
RESEARCH	★ ★ ★ ✬ ☆

BEST TAKEN AS Infusion ✓✓✓
Tincture ✓✓ Capsule ✓
DOSAGE C *(see pp.44–45)*
CAUTIONS Safe at normal dosage. See also pp.42–51.

Cramps and chronic cough Caraway's gently warming and relaxing action within the gut makes it an excellent remedy for soothing digestive problems such as nausea, indigestion, wind, and bloating. Caraway is an effective remedy for colic in children. A common ingredient in cough mixtures, it can be taken to relieve croup and chronic cough. Do not take the essential oil internally unless on professional advice.

a safe remedy for children

FRESH LEAVES

GOTU KOLA

Centella asiatica syn. *Hydrocotyle asiatica*

An ancient medicine, gotu kola has been in continuous use in India for at least 2,000 years. Traditionally thought to strengthen memory and concentration, in the West it is more often used to treat chronic skin disorders and support wound healing.

MEDICINAL USES

Part used Whole plant

Key actions Adaptogen
- Aids cognition • Anti-inflammatory
- Tonic • Wound healer

Wounds and broken tissue Gotu kola has many potential uses, the majority revolving around its ability to promote effective tissue repair and wound healing. It speeds tissue healing and reduces the risk of scar (or keloid) formation, including adhesions. It may be applied as a lotion to the skin to treat, for example, minor burns, psoriasis, and scars.

Cognition support In line with traditional use, current research points to neuroprotective and memory strengthening actions for gotu kola. It is one of a number of herbs that supports cognition and balanced mental activity.

Other uses Taken internally, gotu kola appears to tone and strengthen veins, and is commonly prescribed for leg ulcers, venous insufficiency, and varicose and

KEY INFORMATION

SAFETY	★ ★ ★ ⯪ ☆
TRADITIONAL USE	★ ★ ★ ★ ☆
RESEARCH	★ ★ ★ ☆ ☆

BEST TAKEN AS Tincture ✓✓✓ Tablet ✓✓
Lotion (topical) ✓
DOSAGE B (*see pp.44–45*)
CAUTIONS Can cause allergic reaction. Rarely, may cause gastric irritation. See also pp.42–51.

DRIED HERB

thread veins. As a background treatment, it proves useful in many chronic health problems. Its adaptogenic properties enhance the body's ability to respond to both physical and emotional stress.

A small creeping plant, gotu kola grows in the wild throughout India.

CHICORY

Cichorium intybus

Cultivated for food and medicine for at least 5,000 years, chicory's medicinal benefit has much to do with the strongly bitter taste of its root and leaves. Perhaps best known as a caffeine-free alternative to coffee, chicory root exerts a gentle tonic action throughout the digestive system, relieving wind, bloating, and constipation.

MEDICINAL USES

Part used Root ● Leaves

Key actions Bitter tonic ● Anti-diabetic ● Relieves wind and bloating ● Stimulates appetite

Digestive tonic, supports gut health
Chicory root, like its close relative dandelion (*Taraxacum officinale*), supports digestive function throughout the gut from stomach to colon. It is useful in upper digestive problems such as loss of appetite and weak digestion, and is an excellent remedy for those with windy digestions. A mild laxative, it is particularly safe for young children with constipation. Chicory root contains large amounts of inulin, a prebiotic known to support healthy gut flora, that also appears to inhibit harmful bacteria in the gut.

Other uses Chicory root promotes liver regeneration and healthy pancreatic function, making it a useful remedy for diabetics. It has also been used to treat osteoarthritis. The roasted root is commonly used as a coffee-type drink. Chicory leaf, a cultivated species, is a nutritious salad vegetable.

KEY INFORMATION	
SAFETY	★ ★ ★ ★ ★
TRADITIONAL USE	★ ★ ★ ★ ☆
RESEARCH	★ ★ ☆ ☆ ☆
BEST TAKEN AS	Decoction ✓✓✓
DOSAGE	A
OFTEN USED WITH	Fennel (*Foeniculum vulgare*)
CAUTIONS	None known.

In the summer when in flower, chicory's blue petals act like a clock, opening and then closing at the same time every day.

CINNAMON

Cinnamomum verum

An ancient spice, the inner bark of cinnamon leaves a pleasant and warm taste on the tongue. Its undoubted health benefits are not that well known. Recent research points to an entirely new use for it – cinnamon appears to work with insulin to help stabilize blood sugar levels in the body.

MEDICINAL USES

Parts used Essential oil ● Inner bark

Key actions Antimicrobial ● Aromatic ● Astringent ● Mild stimulant ● Relieves wind

Digestive upsets and colds Cinnamon's warming, stimulant action has made it a favourite remedy for digestive upsets. As an infusion, it helps to soothe wind, bloating, nausea, and indigestion, as well as speed recovery from gastrointestinal infection. It has moderate antibacterial and antifungal activity, and acts against *Helicobacter pylori*, an organism that can cause stomach ulcers. In colds, flu, chest infection, and coughs, cinnamon provides a pleasant treatment that can be safely given to children.

Other uses Cinnamon's ability to stimulate the circulation is often overlooked; taken long term, it strengthens blood flow to the hands and feet, helping those with poor peripheral circulation. It can also be taken on a regular basis – one recommendation is a teaspoon of cinnamon powder at night – to support stable blood sugar levels.

Native to India and Sri Lanka, cinnamon is widely cultivated as a spice and medicine.

INNER BARK

KEY INFORMATION

SAFETY	★ ★ ★ ★ ☆
TRADITIONAL USE	★ ★ ★ ★ ☆
RESEARCH	★ ★ ⯪ ☆ ☆

BEST TAKEN AS Infusion ✓✓✓ Tincture ✓✓ Capsule, powder ✓

DOSAGE C (*see pp.44–45*)

CAUTIONS Rarely, can cause allergic reactions. See also pp.42–51.

GERMAN CHAMOMILE

Chamomilla recutita syn. *Matricaria recutita*

Known more as a pleasant-tasting tea than as a medicine, chamomile provides effective treatment for health problems as diverse as indigestion and acidity, travel sickness, cramps, inflamed skin, and poor sleep. Make sure to use good-quality chamomile to achieve best results.

MEDICINAL USES

Parts used Flower ● Essential oil

Key actions Anti-allergenic
● Anti-inflammatory ● Relaxant
● Relieves spasm ● Soothes digestion
● Wound healer

Digestive and inflammatory conditions
Perhaps the most commonly used European herb, chamomile can be safely taken by babies, children, and adults for all manner of problems affecting the digestive system. From mouth ulcers and stomach ache to colic and looseness, chamomile will soothe inflammation, acidity, and cramps and encourage effective recovery. Regular cups of chamomile tea can make a difference in inflammatory conditions such as gastritis, Crohn's disease, and colitis. For best results, brew chamomile in a teapot or in a cup with the saucer on top because most of the active constituents

essential oil is used to treat skin problems

ESSENTIAL OIL

are present in the steam. Chamomile is an excellent anti-inflammatory when applied topically – use the infusion as a lotion on sore and itchy rashes, grazes, and insect bites and stings. Apply a warm chamomile tea bag to sore or irritated eyes. As a lotion or poultice, the flowers can prove helpful in treating sore nipples and mastitis.

Menstrual pain and cramps
The use of chamomile tea to relieve period pains and cramps predates Roman times. While other herbs may be as good or better in this respect, the ready availability of chamomile makes it an easy herb to select at the time of need.

Nervous tension
Chamomile will serve well in treating other types of cramp, particularly muscle tension resulting from tension or overwork. As a mild sedative and relaxant, it can help to ease anxiety and nervous stress that interferes with normal digestive function, for example, in irritable bowel syndrome. In those prone to nervous tension and cold hands and feet, chamomile combined with ginger (*Zingiber officinale*) – grate fresh root into a teapot – can help if taken regularly.

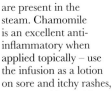

poultice of flowerheads can relieve sore skin

DRIED FLOWERHEADS

KEY INFORMATION

SAFETY	★★★★☆
TRADITIONAL USE	★★★★★
RESEARCH	★★★★☆

BEST TAKEN AS Infusion ✓✓✓
Tincture ✓✓ Essential oil (topically) ✓
DOSAGE A *(see pages 44–45)*
OFTEN USED WITH Meadowsweet (*Filipendula ulmaria*)
CAUTIONS Rarely, can cause allergic reaction. See also pp.42–51.

Children's ailments A first-rate remedy for children, chamomile can be safely given to infants and children from the age of six months upwards. In babies suffering from colic and digestive discomfort, breast-feeding mothers can drink the tea, or add a small cup of chamomile tea to the baby's bath. It soothes fractious and over-tired infants, gently encouraging relaxation and a good night's sleep. Chamomile can be confidently given to children in all gastrointestinal complaints. The tea can bring relief during teething. If taste is a problem, mix the tea with unsweetened apple or blackcurrant juice. Do not take the essential oil internally unless on professional advice.

Found wild throughout Europe and temperate Asia, chamomile has also naturalized in Australia and the USA.

A GARDENER'S DELIGHT

In medieval times, chamomile was thought of as the "plant's physician". It has been stated that nothing contributes as much to the health of a garden as chamomile herbs dispersed about it. A drooping or sickly plant may well recover if chamomile is placed near it.

GARDEN HERB

BLACK COHOSH

Cimicifuga racemosa syn. Actaea racemosa

Recent interest in black cohosh as an alternative to hormone replacement therapy (HRT) has led to a dramatic increase in its popularity, especially as a treatment for menopausal symptoms. A Native American remedy, black cohosh has always been seen as a herb for women's ailments.

MEDICINAL USES

Part used Root

Key actions Anti-inflammatory
● Anti-rheumatic ● Hormonal tonic
● Mild sedative ● Relieves spasm

Menopause Black cohosh has become the herb of choice for treating menopausal symptoms. Although research is divided on its effectiveness, it is well worth trying for relief of menopausal symptoms such as hot flushes, night sweats, disturbed sleep, nervous irritability, and headache. It should be taken for at least 2–3 weeks to see if beneficial effects result. It may prove more effective combined with other suitable herbs, for example sage (*Salvia officinalis*), particularly when these are recommended by a herbal practitioner. Where nervous exhaustion or depression are present, black cohosh should be combined with St John's wort (*Hypericum perforatum*). Black cohosh is taken to treat premenstrual problems

DRIED ROOT

such as irritability, breast tenderness, menstrual pain, and irregular or absent periods. Again, it may prove more effective in combination with other herbs.

Nerve problems Black cohosh has mild sedative activity and can aid nerve-based problems such as pain, chronic headache, migraine, tinnitus, and vertigo. Traditionally, the herb is seen as a "nervine" (calming and strengthening the nervous system), its overall effect on the nervous system being to reduce overactivity and relieve pain.

KEY INFORMATION

SAFETY	★ ★ ★ ★ ☆
TRADITIONAL USE	★ ★ ★ ★ ☆
RESEARCH	★ ★ ★ ☆ ☆

BEST TAKEN AS Tincture ✓✓✓
Capsule ✓✓ Tablet ✓

DOSAGE M, D *(see pp.44–45)*

OFTEN USED WITH Chaste berry (*Vitex agnus-castus*)

CAUTIONS Do not take during pregnancy and while breast-feeding, except on the advice of a herbal or medical practitioner. Larger doses can cause stomach upsets and headache. Rarely, may cause liver damage; avoid in pre-existing liver disease. See also pp.42–51.

FRESH ROOT

Arthritis and rheumatism For muscle pain, arthritis, and rheumatic conditions (especially when associated with the menopause), black cohosh can sometimes prove highly effective in relieving pain and inflammation and improving freedom of movement. As an antispasmodic, the herb eases cramps and restless muscles, and will tend to help lower raised blood pressure.

Cancer Black cohosh has oestrogenic activity within the body, although it appears not to contain oestrogens. Opinion varies about the herb's safety in hormone-dependent cancers such as breast and ovarian cancer. In this situation, seek advice from your herbal practitioner or doctor before starting to take black cohosh.

NATIVE AMERICAN REMEDY

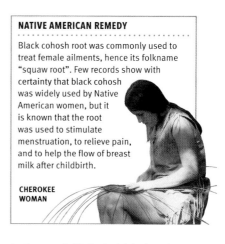

Black cohosh root was commonly used to treat female ailments, hence its folkname "squaw root". Few records show with certainty that black cohosh was widely used by Native American women, but it is known that the root was used to stimulate menstruation, to relieve pain, and to help the flow of breast milk after childbirth.

CHEROKEE WOMAN

Another name for black cohosh is bugbane; its faintly unpleasant smell is said to repel insects.

A remedy for scurvy long before vitamin C was identified, lemon (*Citrus limon*) is a valuable preventative medicine, increasing resistance to infection and helping to maintain good health.

LEMON

Citrus limon

One of the most useful home remedies, lemon works well in a host of common complaints. The traditional drink of lemon and honey can be spiced up with ginger and garlic to make a potent brew for colds, flu, coughs, and digestive disorders. Overall, the fruit improves resistance to infection.

MEDICINAL USES

Parts used Fruit/juice ● Essential oil ● Peel ● Seed

Key actions Antioxidant ● Antiseptic ● Detoxifying agent ● Nutritive

Detoxification Most parts of the lemon can be used medicinally. The juice taken as a cleansing drink, rich in vitamin C and antioxidant bioflavonoids, stimulates liver metabolism and detoxification. Diluted juice of a freshly squeezed lemon makes an excellent pre-breakfast drink. The juice makes an effective mouthwash and gargle; to achieve best results, add a pinch of chilli (*Capsicum* spp.) for mouth ulcers, gingivitis, and sore throat. It also stimulates the appetite, aids digestion, and improves absorption of iron. Its action on the liver means that it helps to reduce the tendency to allergic reaction and promotes the elimination of waste products. Avoid drinking the juice neat, as it is strongly acidic and can dissolve tooth enamel. Brush teeth after drinking lemon juice.

Lemon is said to have originated in India. Its fruit, used as a natural medicine, is harvested in winter when the vitamin C content is at its highest.

Skin disorders Applied topically, lemon juice and essential oil help to counter bacterial and fungal infection and to heal acne spots, chilblains, and insect stings and bites. The peel and pith contain high levels of bioflavonoids and essential oil, making extracts valuable in chronic health problems such as thread veins, varicose veins, and poor peripheral circulation. The seeds, like grapefruit seeds, are antiseptic and, chewed or crushed, can be taken for candidiasis and other fungal problems. On fungally infected nails, apply 1–2 drops of essential oil each day.

juice aids liver metabolism

FRUIT

KEY INFORMATION

SAFETY	★ ★ ★ ★ ★
TRADITIONAL USE	★ ★ ★ ★ ★
RESEARCH	★ ★ ★ ☆ ☆
BEST TAKEN AS	Diluted juice ✓ ✓ ✓
DOSAGE	Juice
OFTEN USED WITH	Ginger (*Zingiber officinale*)
CAUTIONS	Unsuitable in acidic conditions. See also pp.42–51.

COFFEE

Coffea arabica

It is hard to imagine life without coffee culture, but coffee drinking took off in the West only in the 18th century. Coffee's ability to sharpen wit and mental focus, and its effectiveness as a stimulant, has guaranteed its popularity since then.

MEDICINAL USES

Part used Bean (ripe seed)

Key actions Diuretic ● Stimulant

Headache Hard as it may be to think of coffee as a medicine, there is no doubt that it can be put to good medicinal use. Coffee, or caffeine, is a common ingredient in headache and pain-relieving tablets, for example, when used with paracetamol. On its own, coffee can help to clear a foggy head and headache. In moderation, coffee (and caffeine) stimulates alertness, and improves concentration and work rate.

> **KEY INFORMATION**
>
> SAFETY ★ ★ ★ ★ ☆
> TRADITIONAL USE ★ ★ ☆ ☆ ☆
> RESEARCH ★ ★ ★ ☆ ☆
> BEST TAKEN AS Infusion ✓✓✓
> OFTEN USED WITH Cardamom (*Elettaria cardamomum*)
> CAUTIONS Avoid excessive doses, which can cause palpitations. See also pp.42–51.

more flavour and less caffeine than other processes. In natural medicine, coffee is thought to put an unnecessary strain on the body, especially during illness.

POWDERED BEANS

coffee infusion raises level of alertness

COFFEE BEANS

Coffee can, however, cause headache, poor sleep, palpitations, and heart irregularity, although usually only at high levels of intake. Stopping regular coffee intake can lead to headaches lasting for up to 4 days. Its diuretic effect is probably noted by most drinkers. It is best avoided in chronic health problems, especially in long-term weakness and exhaustion.

Forms and flavours Surprisingly, in its percolated form, coffee is the strongest (and least healthy); espresso extracts

Coffee originated in Ethiopia. People in the Middle East used it as a religious drink to help them stay awake during prayers.

CODONOPSIS, DANG SHEN

Codonopsis pilosula

Codonopsis root is sweet tasting and is used in cooking and as a medicine in China. It has many similarities with ginseng, but is milder in action, and can therefore be taken to support stamina and wellbeing in situations such as chronic tiredness, fibromyalgia and long-term stress and tension, where ginseng might prove too strong.

MEDICINAL USES

Part used Root

Key actions Adaptogen ● Tonic ● Anti-anaemic

Adaptogen Codonopsis has a shorter lasting and less fiery action than ginseng (*Panax ginseng*) making it a key restorative remedy for people with chronic health problems, associated with low energy and tiredness. It is particularly helpful for "adrenal overload" with symptoms such as tense neck muscles, headache, and nervous exhaustion. Though poorly researched, codonopsis appears to increase both red and white blood cell levels, making it valuable in anaemia and chronic infection.

Brain remedy One scientific study concluded that codonopsis might help to promote nerve regeneration, and a small clinical trial combining ginkgo (*Ginkgo biloba*) and codonopsis noted a significant improvement in memory and recall.

Other uses Codonopsis appears to help lower high blood pressure, so it is a safe remedy for people with hypertension. In China, the root is regularly taken by nursing mothers to increase milk production and "build strong blood".

A climbing plant, codonopsis has a sweet-tasting root, which is often added to Chinese medicinal soups.

> ### KEY INFORMATION
>
> | | |
> |---|---|
> | SAFETY | ★ ★ ★ ★ ☆ |
> | TRADITIONAL USE | ★ ★ ★ ★ ☆ |
> | RESEARCH | ★ ★ ☆ ☆ ☆ |
>
> **BEST TAKEN AS** Tincture ✓✓✓
> Decoction ✓✓ Tablet ✓
> **DOSAGE** B (*see pp.44–45*)
> **CAUTIONS** None known.
> See also pp.42–51.

DRIED ROOT

MYRRH

Commiphora molmol

A bitter-tasting resin with an ancient provenance, myrrh has been esteemed as a cleansing and antiseptic remedy for many thousands of years. Used in perfumery, it powerfully disinfects tissue with which it comes into contact.

MEDICINAL USES

Part used Resin

Key actions Anti-inflammatory ● Antimicrobial ● Antiseptic ● Astringent ● Stimulant

Skin, digestive problems Strongly antiseptic and astringent – and very unpleasant-tasting – myrrh is used in inflamed and infected conditions affecting the skin and digestive tract. It makes an excellent mouthwash and gargle, either on its own or combined with other herbs such as sage (*Salvia officinalis*). Diluted (or neat) tincture is a valuable first aid remedy to cleanse and disinfect cuts, grazes, and wounds; however, it stings on being applied, especially the neat tincture.

High cholesterol Another antiseptic resin and close relative of myrrh, guggul (*Commiphora mukul*) has a pronounced ability to reduce raised cholesterol levels. Research supports its use (in capsule or tablet form) for this purpose.

RESIN   **CAPSULES**

KEY INFORMATION

SAFETY	★ ★ ★ ☆ ☆
TRADITIONAL USE	★ ★ ★ ★ ★
RESEARCH	★ ★ ☆ ☆ ☆

BEST TAKEN AS Diluted tincture (topical) ✓✓✓
Standardized extract: Tablet ✓✓✓
Capsule ✓✓
DOSAGE M (*see pp.44–45*)
CAUTIONS Do not take during pregnancy and while breast-feeding. Can occasionally cause allergy, skin rashes, digestive disturbance, and headache.
Avoid taking alcoholic extracts. See also pp.42–51.

Native to north-eastern Africa, the trunk and branches of myrrh exude a thick yellow resin that has a strong, aromatic scent.

LILY OF THE VALLEY

Convallaria majalis

A slender plant with fragrant white flowers, lily of the valley acts on a weak heart to improve its functioning. A potent medicinal herb, it can have toxic effects.

MEDICINAL USES

Part used Flowering plant

Key actions Diuretic ● Heart tonic ● Lowers blood pressure

Heart problems The active compounds within lily of the valley are similar to those found in *Digitalis purpurea*, the source of the heart drug digitoxin. Lily of the valley is milder in action and safer to use, being prescribed for functional heart problems.

FLOWERING PLANT

KEY INFORMATION

SAFETY	★ ½ ☆ ☆ ☆
TRADITIONAL USE	★ ★ ★ ★ ☆
RESEARCH	★ ★ ½ ☆ ☆

DOSAGE On prescription or in licensed product only.

CAUTIONS Potentially toxic. Do not take during pregnancy and while breast-feeding. Take only when prescribed by a herbal or medical practitioner. Restricted herb in some countries, including the UK and Australia. See also pp.42–51.

CRATAEVA, VARUNA

Crataeva nurvala

A key remedy for kidney, bladder, and urinary tract problems, crataeva can prove effective in conditions such as cystitis, enlarged prostate, and kidney stones.

MEDICINAL USES

Parts used Root bark ● Stem bark

Key actions Anti-inflammatory ● Bladder tonic

Kidney problems Worth considering for any urinary disturbance, crataeva will help to soothe the urinary tract and bladder, reduce frequency, and together with antiseptic remedies, help to clear infections such as cystitis. It is reputed to help dissolve kidney stones. As with other remedies for the urinary system, crataeva is best taken as a decoction.

DRIED STEM BARK

KEY INFORMATION

SAFETY	★ ★ ★ ★ ☆
TRADITIONAL USE	★ ★ ★ ★ ☆
RESEARCH	★ ★ ★ ☆ ☆

BEST TAKEN AS Decoction ✓✓✓ Tincture ✓✓

DOSAGE B (*see pp.44–45*)

CAUTIONS Do not take during pregnancy and while breast-feeding. See also pp.42–51.

PUMPKIN SEED

Cucurbita pepo

While pumpkin is best known for pie and Halloween masks, its seeds offer well-established nutritional benefits. Full of high-grade essential fatty acids and trace elements, notably zinc, the seeds make an excellent food supplement.

MEDICINAL USES

Parts used Seed ● Seed oil

Key actions Demulcent ● Deworming agent ● Diuretic ● Hormonal agent

Benign prostatic hypertrophy (BPH)
Seeds and seed oil are often used to treat the early stages of enlargement of the prostate gland, or BPH. Recommended by the German Department of Health, the seeds are often effective at relieving BPH symptoms such as difficulty in passing urine. They do not appear to reverse enlargement of the gland.

Other uses Key constituents within the seeds have an oestrogenic activity, so a regular intake may prove helpful in relieving menopausal symptoms. The seeds make an effective treatment for worms, notably tape worm, and can be safely used by children and adults, including during pregnancy. Large quantities need to be taken; seek professional advice.

KEY INFORMATION
SAFETY ★ ★ ★ ★ ☆
TRADITIONAL USE ★ ★ ★ ★ ☆
RESEARCH ★ ★ ★ ☆ ☆
BEST TAKEN AS Seed ✓✓✓
DOSAGE 10g a day
OFTEN USED WITH Saw palmetto (*Serenoa serrulata*)
CAUTIONS None known. See also pp.42–51.

seeds are an effective deworming agent

DRIED SEEDS

Pumpkin seeds are a good dietary source of iron, zinc, and selenium.

fruit pulp is used as a poultice for burns

HAWTHORN

Crataegus spp.

Regarded by herbalists as a "food for the heart", hawthorn is one of the most scientifically validated of all herbal medicines, exerting specific benefit on the heart. Both berries and flowering tops improve blood flow through the coronary arteries to the heart.

MEDICINAL USES

Parts used Flowering top ● Leaf ● Fruit

Key actions Antioxidant ● Heart tonic ● Lowers blood pressure ● Relaxes blood vessels

Coronary diseases Hawthorn is not a cure-all for heart and circulatory disorders, but if used carefully and when taken long term, it will lead to improvement in cardiovascular health. Hawthorn works directly on the heart to slow its rate, improve oxygen uptake, and increase its pumping efficiency. Specific indications include palpitations and heart irregularity, mild angina, and early signs of heart weakness. Evidence from clinical trials supports hawthorn's use in the early stages of heart disease. In such situations, and especially where prescribed medicines are

Bright red berries appear in autumn. Although sour in taste, hawthorn berries, like several other red berries, were formerly used to make desserts.

being taken, seek professional advice from a herbal or medical practitioner before starting treatment with the herb.

High and low blood pressure The berries, flowers, and leaves contain high levels of procyanidins, flavonoid compounds which have a strong antioxidant activity that supports

berries are used to improve cardiovascular health

DRIED BERRIES

"HEART" OF THE MATTER

In the 19th century, an Irish physician called Dr Green became famous for his secret remedy for heart disease. After his death it was revealed that his "cure" was actually a tincture made of hawthorn berries.

19TH-CENTURY PHYSICIAN

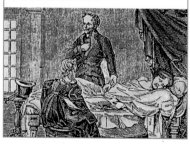

DRIED FLOWERING TOPS

healthy circulation. Until recently, the berries were preferred for treating high blood pressure, but the leaves and flowering tops have been shown to have the greatest concentration of procyanidins and are now more commonly used to treat high blood pressure. It is thought that the leaves help to normalize blood pressure by slowing the heart rate and lowering blood pressure in overactive states, and stimulating the heart rate and raising blood pressure in underactive states.

Circulation Hawthorn contains several substances that actively support the health of the arteries, and may be used to treat or prevent circulatory problems such as atherosclerosis and poor peripheral circulation. Other conditions that may benefit from the herb include intermittent claudication and Raynaud's phenomenon (poor circulation to hands and feet).

KEY INFORMATION

SAFETY	★ ★ ★ ★ ☆
TRADITIONAL USE	★ ★ ★ ★ ☆
RESEARCH	★ ★ ★ ★ ☆

BEST TAKEN AS Standardized extract: Tincture ✓✓✓ Tablet ✓✓ Capsule ✓
DOSAGE M, C (*see pp.44–45*)
Berry: Not less than 4mg/ml of procyanidins
Leaf: Not less than 10mg/ml of procyanidins
OFTEN USED WITH Yarrow
(*Achillea millefolium*)
CAUTIONS Interaction with prescribed medicines can occur. Seek advice from a herbal or medical practitioner if taking prescribed medicines, especially for high blood pressure and heart disorders. Rarely, can cause mild symptoms such as headache and digestive upset. See also pp.42–51.

Hawthorn flowers, or "May blossom", were traditionally used in May Day celebrations throughout Europe and have a long association with fertility.

TURMERIC

Curcuma longa

A key component of curry mixtures, turmeric's golden-yellow colour is familiar to all who eat Indian food. Turmeric root has traditionally been taken to heal allergic and inflammatory conditions and research has established that it has extensive health benefits, due in particular to its strong anti-inflammatory activity.

MEDICINAL USES

Part used Root

Key actions Anti-inflammatory
● Antioxidant ● Protects liver
● Stimulates bile flow

Detoxicant Turmeric promotes healthy function within the upper digestive system as a whole, countering infection and inflammation within the stomach and small intestine. At the same time, it acts to protect the liver from toxic damage and stimulates bile flow.

Cancer prevention Turmeric has a role to play in many chronic health problems and is increasingly being used as a cancer-preventative. More controversially, turmeric may be a valuable supplement to

take in order to support good health where cancer has been diagnosed. In this case, take the herb only on the recommendation of a qualified herbal practitioner or doctor.

Other uses The root's marked antioxidant activity means that it has a role to play in much chronic illness. Recent research suggests possible benefit in conditions as diverse as indigestion and nausea, gastritis, peptic ulcer, liver disorders, high blood cholesterol levels, arthritis, and inflammatory autoimmune problems such as rheumatoid arthritis and Crohn's disease. Turmeric also has antifungal and antibacterial activity and can help treat candidiasis. The powder can be made into a paste and applied to infected areas of the skin.

> **KEY INFORMATION**
>
> SAFETY   ★ ★ ★ ★ ☆
> TRADITIONAL USE   ★ ★ ★ ★ ☆
> RESEARCH   ★ ★ ★ ★ ☆
> BEST TAKEN AS Tablet ✓✓✓ Powder ✓✓
> Tincture ✓
> **DOSAGE** B *(see pp.44–45)*
> **OFTEN USED WITH** Ginger
> (*Zingiber officinale*)
> **CAUTIONS** If taking blood-thinning medication or if gallstones are present, take only on the advice of a herbal or medical practitioner. See also pp.42–51.

turmeric is used as
a remedy for gastritis

TURMERIC POWDER

Native to India and southern Asia, turmeric has been used in both Ayurvedic and ancient Chinese herbal medicine to treat liver problems, including jaundice.

GLOBE ARTICHOKE

Cynara scolymus

The flowerheads of globe artichoke make a tasty vegetable dish and, like the leaves, have a tonic action on the liver and digestion, stimulating appetite and detoxification. However, the leaves are mostly used in medicine, with substantial evidence to prove that they lower cholesterol levels.

MEDICINAL USES

Parts used Flowerhead (food) ● Leaf (medicine)

Key actions Antioxidant ● Bitter and digestive tonic ● Lowers cholesterol levels ● Protects the liver

Liver and kidney problems Another herb where recent research has found new uses, globe artichoke remains a key herb for strengthening liver and kidney function, thus supporting detoxification in chronic conditions such as arthritis, gout, and liver disease.

High cholesterol Clinical trials over the last 30 years have found that globe artichoke leaf lowers cholesterol and triglyceride levels, while high density lipoprotein (HDL) levels tend to increase.

The improvement in cholesterol levels varied from 5 to 45 per cent, with a daily dose of 7g equivalent of dried leaf. It should be taken for some months to achieve best results. Patients also reported significant relief from symptoms such as nausea, vomiting, abdominal pain, flatulence, and constipation. Following the outcome of these trials, globe artichoke is now commonly taken to treat irritable bowel syndrome and related symptoms such as bloating, abdominal distension, and alternating constipation and diarrhoea.

Globe artichoke is a traditional treatment for jaundice and kidney stones.

DRIED LEAVES

KEY INFORMATION

SAFETY	★ ★ ★ ★ ★
TRADITIONAL USE	★ ★ ★ ★ ☆
RESEARCH	★ ★ ★ ⯨ ☆

BEST TAKEN AS Tablet ✓✓✓ Tincture ✓✓
DOSAGE A (*see pp.44–45*)
OFTEN USED WITH Turmeric (*Curcuma longa*)
CAUTIONS If gallstones are present, take only on the advice of a herbal or medical practitioner. Rarely, can cause allergic reactions and wind and bloating. See also pp.42–51.

Native to the Mediterranean, globe artichoke (*Cynara scolymus*) is used both as food and medicine. Its strongly antioxidant and liver-protective leaves are particularly effective.

WILD YAM

Dioscorea villosa

Increasingly taken as a remedy to relieve menopausal symptoms, wild yam has traditionally been used to ease cramps and muscle pain, especially menstrual pain and colic, throughout the body. A further key use has been in the treatment of inflammatory arthritis, including rheumatoid arthritis.

MEDICINAL USES

Part used Root

Key actions Anti-inflammatory ● Oestrogenic ● Relieves spasms

Cramps and pains Wild yam can bring relief wherever cramping pain or over-tensed muscles are the main symptoms. In such a situation, wild yam's combination of antispasmodic and anti-inflammatory activity can help to soothe problems as diverse as intestinal cramps, gall bladder pain, menstrual and ovarian pain, and muscle spasm resulting from chronic inflammation. In many cases, the best results will be obtained by combining wild yam with other anti-inflammatory or muscle-relaxant remedies, particularly cramp bark (*Viburnum opulus*). In both osteoarthritis and rheumatoid arthritis, wild yam combines effectively with anti-inflammatories such as devil's claw (*Harpagophytum procumbens*) and willow bark (*Salix alba*).

Menopausal symptoms Wild yam is best known, and most commonly taken, for the relief of menopausal symptoms. Given the herb's undoubted hormonal

TINCTURE

activity, there are good reasons for thinking that it can be taken to improve symptoms such as hot flushes, night sweats, and poor sleep, though it is probable that the steroid compounds in wild yam are not converted to the active hormones in the human body that would thus make it effective for menopausal symptoms; perhaps there is another mechanism at work. Most experts

KEY INFORMATION

SAFETY	★ ★ ★ ★ ⯨
TRADITIONAL USE	★ ★ ★ ⯨ ☆
RESEARCH	★ ★ ☆ ☆ ☆

BEST TAKEN AS Tincture ✓✓✓
Capsule ✓✓ Tablet ✓
DOSAGE B (*see pp.44–45*)
OFTEN USED WITH Black cohosh (*Cimicifuga racemosa*)
CAUTIONS Can cause irritation within the digestive tract, usually only with excessive dosage. See also pp.42–51.

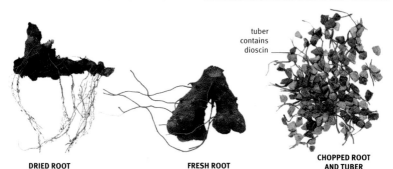

tuber contains dioscin

DRIED ROOT

FRESH ROOT

CHOPPED ROOT AND TUBER

Native to the southern USA, Mexico, and Central America, wild yam was known as colic root and rheumatism root, indicating its use by early settlers.

recommend that for best results, the extract should be taken for several weeks to see if it alleviates symptoms. Wild yam combines well with black cohosh (*Cimicifuga racemosa*) for menopausal and rheumatic problems.

Natural progesterone cream Wild yam natural progesterone cream, applied to the skin rather than taken internally, has received much publicity as a treatment for menopausal problems. A number of clinical studies have failed to find any benefit in relieving symptoms in menopausal women, though there are undoubtedly women who have experienced relief of symptoms with the cream. It is perhaps misleading to describe the product as "natural", as several laboratory processes are required in order to convert the steroidal compounds present in wild yam into progesterone. No plant has yet been found that contains progesterone. That being said, it is the case that hormones processed from natural sources are more readily used by the body than those that are produced synthetically.

Other uses Wild yam has been used within Native American traditions to help prevent miscarriage in the later stages of pregnancy and to relieve pain during childbirth. In keeping with its traditional name of colic root, wild yam makes a useful treatment for irritable bowel and diverticulitis, especially if combined with slippery elm (*Ulmus rubra*).

HORMONES FROM WILD YAM

Wild (and other) yams contain high levels of steroidal compounds, including dioscin, that have an oestrogenic activity. It was the original plant source of diosgenin produced by Japanese scientists in 1936. Diosgenin was synthesized in the laboratory into steroid hormones, and eventually led to the creation of the birth control pill.

STEROID MOLECULE

ECHINACEA

Echinacea angustifolia, E. purpurea, E. pallida

A plant from the plains of North America, echinacea is thought to powerfully stimulate the body's ability to resist infection and counter toxicity. Taken mainly as a treatment or preventative for common cold, flu, and viral infections, the herb also helps to heal skin disorders such as eczema and acne.

MEDICINAL USES

Parts used   Whole plant

Key actions   Antibacterial ● Anti-viral ● Blood cleanser ● Immune-enhancing ● Wound healer

DRIED ROOT

Colds, flu, viral and bacterial infection

Echinacea – as tincture, tablet, or capsule – is frequently taken to speed recovery from colds, sore throats, and chest infections. The herb is also known to enhance immune resistance in those prone to recurrent colds or herpes outbreaks or where flu-like symptoms linger. Echinacea combines well with the flower or berry of elder (*Sambucus nigra*) in this context. The diluted tincture makes a mouthwash or gargle, and can be used to wash infected skin rashes and wounds. Bacterial infections such as sinusitis, tonsillitis, and chronic bronchitis can be self-treated with echinacea,

preferably in combination with remedies such as garlic (*Allium sativum*) and golden seal (*Hydrastis canadensis*), but a fever of 39°C (102°F) or over indicates the need to seek professional advice. Although the evidence is reasonably good, there is still some debate on echinacea's effectiveness in treating and preventing infection. This

root is used to boost the immune system

FRESH ROOT

KEY INFORMATION

SAFETY	★ ★ ★ ★ ☆
TRADITIONAL USE	★ ★ ★ ★ ☆
RESEARCH	★ ★ ★ ☆ ☆

BEST TAKEN AS   Tincture ✓✓✓ Tablet ✓✓ Capsule ✓

DOSAGE B *(see pp.44–45)*

OFTEN USED WITH   Golden seal (*Hydrastis canadensis*)

CAUTIONS   Can cause allergic reactions. If taking prescribed medication, seek advice from a herbal or medical practitioner. See also pp.41–52.

may be partly because the dosage used in some clinical trials was too low, or the wrong plant parts or species were studied. The quality and form of echinacea products are important – *E. angustifolia* or extracts of the freshly pressed juice of the above-ground parts of *E. purpurea* are thought to have the strongest medicinal activity.

Immune support and detoxification

Echinacea stimulates non-specific immunity, increasing the number and activity of white blood cells. This makes it a front-line remedy wherever the immune system is overburdened by chronic infection or toxic residues such as in swollen lymph glands, recurrent boils, chronic dull headache, or sore throat. Used carefully – and this means in conjunction with a herbal practitioner – it helps in cleansing the lymph system, supporting resistance to underlying infection such as fungal problems, and improving overall vitality. Echinacea is not suitable for self-treatment in autoimmune disorders or HIV infection.

SNAKE ROOT

Knowledge of echinacea's medicinal value comes down to us from the experience of Native Americans. Used traditionally as a treatment for snake bite (hence its other common name, "snake root"), echinacea has been used to prevent septic infection in wounds and as a remedy for toothache, sore throat, and rabies.

RATTLE SNAKE

Three echinacea species are used medicinally. All are threatened in the wild, but *Echinacea purpurea* is cultivated widely in the USA and Europe.

A Native American remedy for septic conditions and snake bites, echinacea (*Echinacea* spp.) is today the most important immune tonic in Western herbal medicine.

CARDAMOM

Elettaria cardamomum

A well-known Indian condiment, cardamom has a warm, slightly spicy taste, and can be added to sweet and savoury dishes alike. Its seeds combine well with other remedies to improve flavour and to soothe an upset digestive system. They also add zest to coffee, making a subtler stimulant than coffee alone.

MEDICINAL USES

Parts used Seed ● Essential oil

Key actions Mild stimulant ● Relieves wind ● Soothes digestion ● Tonic

Wind, bloating, digestive ill health
Cardamom's main therapeutic use lies in easing discomfort within the upper digestive system, making it a valuable digestive remedy. Mildly warming and analgesic, its oil relieves colic and wind, and helps to settle nausea, griping, and indigestion. It combines well with chamomile (*Chamomilla recutita*). Use the crushed seeds or the tincture; take the essential oil internally only on professional advice.

KEY INFORMATION

SAFETY	★ ★ ★ ★ ★
TRADITIONAL USE	★ ★ ★ ⯪ ☆
RESEARCH	★ ★ ⯪ ☆ ☆

BEST TAKEN AS Crushed seed ✓✓✓ Tincture ✓✓
DOSAGE C *(see pp.44–45)*
OFTEN USED WITH Ginger (*Zingiber officinale*)
CAUTIONS None known.
See also pp.42–51.

seeds are an effective remedy for bad breath

SEEDPODS

Throat and chest problems
The seeds' warming and slightly antiseptic action extends to the throat and chest, making it a good addition to gargles for sore throat, and in chest problems such as asthma and bronchitis.

High blood pressure remedy
Recent clinical research in India found that 3g of cardamom seeds a day taken for 3 months by people with high blood pressure produced consistently lower blood pressure readings. Cardamom is definitely worth considering as part of a natural approach to treating high blood pressure.

TINCTURE

An aromatic herb, cardamom is one of the oldest spices in the world. Apart from its medicinal value, it was also used in perfumes by the ancient Egyptians.

CLOVE

Eugenia caryophyllata syn. *Syzygium aromaticum*

Originally from the Spice Islands of Indonesia, cloves hold a revered place in oriental herbal medicine and cuisine. A potent antiseptic, cloves added to food help to prevent food-borne infection and food poisoning. The essential oil is an excellent first aid remedy for mouth ulcers, toothache, and nerve pain in general.

MEDICINAL USES

Parts used Essential oil
● Flower bud (clove)

Key actions Analgesic ● Anti-emetic ● Antioxidant ● Antiseptic ● Astringent ● Stimulant

Toothache and nerve pain A clove tucked in the mouth, or one drop of essential oil placed on cotton wool and plugged into a tooth, is a tried and trusted remedy for toothache. It should be used sparingly and the oil should not be placed on the gum. The diluted oil (maximum 3 per cent concentration) may also be applied to the skin to relieve nerve pain elsewhere in the body, such as in shingles.

Digestive upset and irritable bowel syndrome With a positive action on the stomach, small doses of clove powder or tincture make a useful remedy in conditions such as nausea, indigestion, wind, and bloating. Cloves can bring relief in gastroenteritis and diarrhoea and can counter infection. Mildly anaesthetic, it is worth trying in irritable bowel syndrome, where it may reduce nerve sensitivity within the gut, easing spasms and urgency.

Grown extensively in Tanzania and Madagascar, cloves are originally from the Molucca Islands in Indonesia and the southern Philippines.

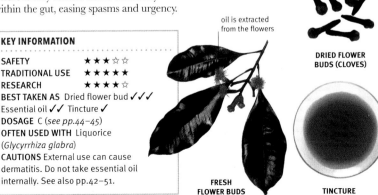

oil is extracted from the flowers

DRIED FLOWER BUDS (CLOVES)

FRESH FLOWER BUDS

TINCTURE

KEY INFORMATION

SAFETY	★★★☆☆
TRADITIONAL USE	★★★★★
RESEARCH	★★★★☆

BEST TAKEN AS Dried flower bud ✓✓✓
Essential oil ✓✓ Tincture ✓
DOSAGE C (*see pp.44–45*)
OFTEN USED WITH Liquorice (*Glycyrrhiza glabra*)
CAUTIONS External use can cause dermatitis. Do not take essential oil internally. See also pp.42–51.

SIBERIAN GINSENG

Eleutherococcus senticosus

First brought into prominence in the West by Soviet research, Siberian ginseng has been routinely given to Russian cosmonauts to aid endurance in space. Similar to Korean ginseng, Siberian ginseng improves the ability to adapt to all kinds of stress, physical and mental.

MEDICINAL USES

Part used Root

Key actions Adaptogen ● Immune-stimulant ● Tonic

Low stamina and endurance Siberian ginseng is a key adaptogen to enhance physical performance and stamina. Research confirms its action in supporting physical endurance and the ability to cope with increased levels of stress and strain, whether these are caused by physical, chemical, environmental, or emotional factors. Its range of indications is therefore very wide, and includes overwork, jetlag, hard physical work, extremes of heat or cold, exposure to radiation, and any situation involving prolonged effort (although not in cases of high blood pressure). For those preparing for exams,

TABLETS

Siberian ginseng combines well with rosemary (*Rosmarinus officinalis*) and ginkgo (*Ginkgo biloba*). The standard advice is to take the herb for up to 6 weeks at a time, and to avoid caffeine, which is thought to undermine the herb's stamina-building effect.

Tiredness and exhaustion Where vitality is low, Siberian ginseng will usually help to improve energy levels, especially in older adults. Although not all states of tiredness and exhaustion will respond to the herb's tonic qualities, many will, notably where adrenal or thyroid gland function is being compromised by overactivity and lack of opportunity to relax and rest. In severe exhaustion, start with a very low dose and slowly increase. It combines well with golden root (*Rhodiola rosea*) and withania (*Withania somnifera*) where ongoing demands make adequate rest hard to come by.

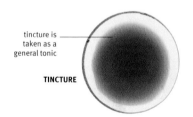

tincture is taken as a general tonic

TINCTURE

root is taken to improve mental and physical resilience

FRESH ROOT

KEY INFORMATION

SAFETY ★★★★✩
TRADITIONAL USE ★★★★☆
RESEARCH ★★★✩☆
BEST TAKEN AS Standardized extract:
Tincture ✓✓✓ Tablet ✓✓
DOSAGE C *(see pp.44–45)*
CAUTIONS Not advisable in high blood pressure. May interact with other medication. See also pp.42–51.

Chronic illness Siberian ginseng can provide much-needed support in times of poor health. Usually in combination with other immune-stimulant remedies, its immune-enhancing and tonic properties make it suitable for chronic viral infections such as shingles, glandular fever, and chronic fatigue syndrome. Even in severe illness, Siberian ginseng can contribute to an improved quality of life. In debilitated states and in convalescence, Siberian ginseng is best taken for several months at a low to medium dosage. It is most likely to strengthen vitality when any improvement in energy levels is used to nurture a return to good health, and not spent on meeting external demands such as work.

Cancer A valuable remedy to aid recovery after surgery or radiotherapy, it plays a role in supporting people with cancer where immune resistance needs to be strengthened, and the ability to tolerate chemotherapy improved. Professional advice must be sought in such cases, not least because the herb can interact with other medication.

Similar in many ways to Korean ginseng (*Panax ginseng*), Siberian ginseng helps to maintain performance and a sense of well-being when one is under long-term stress.

ADAPTOGENS

The term "adaptogen" was coined by Soviet scientists Lazarev and Brekhman in the 1960s to describe herbal medicines that enabled the body and mind to respond more effectively to stress of all kinds. Following research into Siberian ginseng, Russian cosmonauts have used extracts of the herb as an adaptogen to aid stamina and improve the ability to cope with weightlessness.

COSMONAUT'S MEDAL

Californian poppy (*Eschscholzia californica*) is a native of grassy areas of western North America. Its gently sedative and analgesic qualities make it a valuable remedy for anxiety and pain.

CALIFORNIAN POPPY

Eschscholzia californica

Although a close relative of the opium poppy, Californian poppy is safe and non-addictive, and makes a gentle and effective sedative for children.

MEDICINAL USES

Part used Whole plant

Key actions Mild analgesic ● Mild sedative ● Relaxant

Sleep difficulties Best taken in the evening for short-term sleep disturbance, Californian poppy improves sleep quality and can be helpful for nightmares and

FLOWERS

bed-wetting. It combines well with passion flower (*Passiflora incarnata*).

Children's remedy Californian poppy soothes overactivity and also benefits conditions involving pain or anxiety such as headache, migraine, and irritability.

KEY INFORMATION	
SAFETY	★★★★⯪
TRADITIONAL USE	★★★⯪☆
RESEARCH	★★★☆☆
DOSAGE C *(see pp.44–45)*	
CAUTIONS None known within normal dosage, but may have additive effects with alcohol or other sedative herbs/medication. Long-term use not advised. See also pp.42–51.	

EUCALYPTUS

Eucalyptus globulus

A key aboriginal remedy of Australia, eucalyptus can be used to treat everything from colds and chest infection to skin conditions and fever.

MEDICINAL USES

Parts used Leaf ● Essential oil

KEY INFORMATION	
SAFETY	★★★★⯪
TRADITIONAL USE	★★★★★
RESEARCH	★★★★☆
BEST TAKEN AS Manufactured products ✓✓✓	
DOSAGE M *(see pp.44–45)*	
CAUTIONS Not suitable for children under 5. Do not take essential oil internally, except on professional advice. Overuse of the oil topically may be dangerous. See also pp.42–51.	

Key actions Antiseptic ● Expectorant

Coughs and colds The oil is used in many over-the-counter preparations. Main uses include cold symptoms with or without feverishness, nasal and sinus congestion, sore throat, and phlegmy cough.

Skin problems Apply an infusion of the leaves or diluted oil (maximum 5 per cent concentration) to insect bites and fungal skin conditions.

FLOWERING STEM

EYEBRIGHT

Euphrasia officinalis

This herb may have gained its name from its flowers, thought to resemble the human eye. More probably, experience taught that the herb was good for the eyes, hence eyebright. Although little researched, it contains constituents known to have a tonic effect on mucous membranes, including those of the eyes.

MEDICINAL USES

Part used Whole herb

Key actions Anticatarrhal ● Anti-inflammatory ● Astringent

Eye disorders In western herbal medicine, eyebright is a specific remedy for common eye problems such as conjunctivitis ("red eye") and blepharitis (infection of the eyelid). Taken internally, it is thought to heal the surface of the eye, relieving inflammation and drying up excessive watering. Well filtered eyebright infusion can also be applied as a lotion to a partially closed eye.

DRIED LEAVES

leaves are used to relieve inflammation of the eye

Traditionally, the herb has been taken to improve vision.

Catarrh and hay fever Routinely taken by hay fever sufferers each summer, eyebright helps to control allergic symptoms such as sneezing, itchy eyes, and copious watery mucus. Similar conditions that affect the ear, nose, and throat are also likely to benefit from its use.

KEY INFORMATION

SAFETY ★★★★☆
TRADITIONAL USE ★★★★☆
RESEARCH ★★☆☆☆
BEST TAKEN AS Infusion ✓✓✓
Tincture ✓✓ Tablet, capsule ✓
DOSAGE B (*see pp.44–45*)
OFTEN USED WITH Elder (*Sambucus nigra*)
CAUTIONS Possible eye irritation when used topically. See also pp.42–51.

Poorly researched, eyebright is regarded by herbalists as a key remedy for mucous membranes of the eyes, ears, and sinuses.

MEADOWSWEET

Filipendula ulmaria

Aspirin-like substances were first isolated in the 19th century from meadowsweet. The herb has some of the characteristic properties of aspirin, notably a mild anti-inflammatory activity, but unlike aspirin, meadowsweet is a key remedy for healing acid-related problems such as heartburn and gastric ulcer.

MEDICINAL USES

Part used Flowering top

Key actions Antacid ● Anti-inflammatory ● Anti-rheumatic ● Astringent

Native to Europe, meadowsweet flourishes in damp places.

Acid indigestion, gastric ulcer, irritable bowel One of the best remedies for acidic digestive problems, meadowsweet promotes stomach repair while controlling acid release. Taken symptomatically, meadowsweet tea relieves mild heartburn or acid reflux, though for best results, meadowsweet may need to be taken long term. The herb's astringent, binding quality makes it a useful treatment for chronic diarrhoea and irritable bowel.

Rheumatic aches and pains, fibromyalgia Meadowsweet can bring relief to stiff, sore, and aching muscles and joints, soothing inflammation and stimulating clearance of acid residues. Where symptoms get worse on waking or sitting for long periods, combine it with celery seed (*Apium graveolens*) in order to ease inflammation and aid free movement.

DRIED FLOWERS

KEY INFORMATION

SAFETY	★ ★ ★ ★ ☆
TRADITIONAL USE	★ ★ ★ ☆ ☆
RESEARCH	★ ★ ☆ ☆ ☆

BEST TAKEN AS Infusion ✓✓✓
Tincture ✓✓
DOSAGE A *(see pp.44–45)*
OFTEN USED WITH Celery seed (*Apium graveolens*)
CAUTIONS Contains aspirin-like substances; if allergic to aspirin, do not use. Can cause gastrointestinal upset. See also pp.42–51.

FENNEL

Foeniculum vulgare

Fennel tea's pleasant flavour and aroma make it a refreshing drink with marked benefits for digestive health. Safe for children, it gently warms and stimulates appetite and digestion, in the process relieving colic, wind, and bloating. Traditional use is wide-ranging, from relieving menstrual pain to shortness of breath.

MEDICINAL USES

Parts used Seed ● Essential oil

Key actions Eases wind and cough ● Improves appetite and digestion ● Increases breast milk

Indigestion, wind, bloating, colic
Fennel's pleasant taste makes it a popular remedy for upper digestive problems. It relieves griping and abdominal bloating, clears trapped wind, and improves appetite. The diluted tea can be given to young children to relieve colic and teething pain, and is also known to reduce food cravings.

Sore throat, cough, catarrh Fennel tea makes an effective gargle, soothing mucous membranes and relieving cough.

KEY INFORMATION	
SAFETY	★ ★ ★ ★ ☆
TRADITIONAL USE	★ ★ ★ ★ ⯪
RESEARCH	★ ★ ★ ⯪ ☆
BEST TAKEN AS	Infusion ✓ ✓ ✓
DOSAGE	C (*see pp.44–45*)
OFTEN USED WITH	Peppermint (*Mentha x piperita*)
CAUTIONS	Can cause allergic reaction. Do not exceed recommended dose or take for long periods of time. See also pp.42–51.

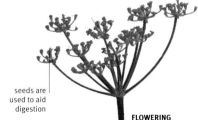

DRIED SEEDS

Hormonal benefits
Fennel increases the production of breast milk and may be taken to start or to maintain a sufficient flow of milk. Taken over a few months, fennel can help to improve menstrual regularity and will tend to reduce menstrual cramps. The seeds have a long-standing reputation as an aid to weight loss and can be added to the diet when trying to lose weight.

seeds are used to aid digestion

FLOWERING HERB

As early as Roman times, fennel was thought to control obesity; a tea made from the seeds normalizes the appetite.

The umbels of golden-yellow flowers and dark green, soft feathery leaves make fennel (*Foeniculum vulgare*) easily recognizable in the pastures and gardens where it grows.

BLADDERWRACK, KELP

Fucus vesiculosus

A cool-water sea vegetable, bladderwrack absorbs large quantities of minerals from the sea. Containing significant levels of iodine – the mineral most responsible for stimulating thyroid gland function – bladderwrack has traditionally been used as part of a weight-loss regime.

MEDICINAL USES

Part used Plant (thallus)

Key actions Demulcent ● Nutritive ● Stimulates thyroid gland ● Supports weight loss

Weight loss and underactive thyroid gland Although as yet unconfirmed by research, anecdotal evidence suggests that bladderwrack is an effective supplement in weight-loss regimes where the thyroid gland is underactive. The herb is a specific for low thyroid function – a condition that causes low vitality, depressed mood and mental function, weight gain, and sensitivity to cold. It can quickly help to reverse symptoms where the thyroid gland is only mildly underactive, or where iodine deficiency

> ### KEY INFORMATION
>
> SAFETY   ★ ★ ★ ☆
> TRADITIONAL USE   ★ ★ ★ ☆
> RESEARCH   ★ ★⯪ ☆ ☆
> BEST TAKEN AS Capsule, tablet ✓✓✓
> DOSAGE B *(see pp.44–45)*
> CAUTIONS Do not take during pregnancy, while breast-feeding, or when thyroid is overactive. See also pp.42–51.

is the principal problem. Treating thyroid disorders can be very complicated and it is advisable to seek professional advice for an apparently underactive thyroid.

Other uses Containing a wide range of minerals, including iodine, silicon, zinc, and copper, bladderwrack is a useful mineral supplement in its own right. Bladderwrack also supports the elasticity and overall health of arteries. With a none-too-pleasant taste, bladderwrack is best taken in capsule or tablet form. It will improve arthritic and rheumatic symptoms where these are linked to low thyroid function. While it is safe at normal dosage, bladderwrack can have adverse effects on the thyroid gland. It can also be contaminated by heavy metals, so quality control is essential.

seaweed extracts can aid weight loss

DRIED SEAWEED

Native to North Atlantic shores and the western Mediterranean sea, bladderwrack is a brownish-green sea plant rich in iodine and silicon.

REISHI, LING-ZHI

Ganoderma lucidum

Used for over 4000 years in China, Japan, and Korea, reishi mushroom has traditionally been taken as a calming tonic in old age. It has also been used for many age-related health problems, including heart and liver disease, inflammatory arthritis, and cancer.

MEDICINAL USES

Part used Fungus

Key actions Antioxidant ● Anti-cancer ● Heart tonic ● Protects liver

Anti-cancer remedy Scientific studies are beginning to validate some of reishi's traditional uses: extracts have been shown to stimulate immune function and promote anti-tumour activity. A safe remedy, reishi is increasingly being used as a supplementary treatment in cancer, especially during chemotherapy. In this case, reishi supports both immune and liver function. In cancer and other serious illnesses, take only on professional advice. While it is thought to have low toxicity, long-term use (more than three months) may cause side effects.

Other uses Reishi's complex action upon the immune system means that it can prove helpful wherever immune function is compromised. A range of chronic health problems such as candidiasis, chronic fatigue, glandular fever, and HIV can all potentially benefit from its use. Reishi is most likely to prove effective when it is used in combination with other treatments – herbal or conventional, as required. Reishi also possesses antiallergenic properties and can be taken to treat allergic conditions such as bronchial asthma. It is also taken by Chinese mountaineers to help prevent altitude sickness.

reishi protects the liver from toxic damage

DRIED MUSHROOMS

Found growing wild on tree trunks or stumps in coastal China, reishi mushrooms are now widely cultivated in North America, Japan, and Korea.

KEY INFORMATION

SAFETY ★★★★⯪
TRADITIONAL USE ★★★★☆
RESEARCH ★★★☆☆
BEST TAKEN AS Standardized extract ✓✓✓ Tincture ✓✓
DOSAGE M (*see pp.44–45*)
OFTEN USED WITH Shiitake (*Lentinus edodes*)
CAUTIONS Can cause allergic reactions. See also pp.42–51.

GENTIAN

Gentiana lutea

Containing some of the most bitter-tasting substances on the planet, gentian is classified in herbal medicine as a "pure" bitter.

MEDICINAL USES

Part used Root

Key actions Bitter ● Digestive tonic

Weak digestion Like all bitters, gentian works to stimulate and strengthen digestive activity. A few drops of gentian tincture is a specific for weak digestion in elderly people and convalescing patients. It stimulates liver and pancreatic function and the release of stomach acid. Increased digestive juices also lead to a healthier appetite.

KEY INFORMATION

SAFETY	★ ★ ★ ★ ✬
TRADITIONAL USE	★ ★ ★ ★ ★
RESEARCH	★ ★ ★ ★ ☆
BEST TAKEN AS	Tincture ✓ ✓ ✓
DOSAGE	D *(see pp.44–45)*
CAUTIONS	None known. See also pp. 42–51.

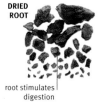
DRIED ROOT

root stimulates digestion

FLOWERS

GYMNEMA, GURMAR

Gymnema sylvestre

Long used in India to treat diabetes and poor sugar tolerance, gymnema stimulates insulin production and may help pancreatic function.

MEDICINAL USES

Part used Leaf

Key actions Antidiabetic ● Lowers blood sugar levels

Sugar craving and poor sugar tolerance Gymnema's traditional name of "gurmar" means "sugar destroyer", gained because it neutralizes the sweet taste buds in the tongue and tackles sugar cravings. Take 20–50 drops of tincture on the tongue every 3 hours. Larger doses are required to improve pre-diabetic states and diabetes, which should be treated only on the advice of a health professional.

KEY INFORMATION

SAFETY	★ ★ ★ ★ ☆
TRADITIONAL USE	★ ★ ★ ★ ☆
RESEARCH	★ ★ ★ ✬ ☆
BEST TAKEN AS	Tincture ✓ ✓ ✓
DOSAGE	D *(see pp.44–45)*
CAUTIONS	Will interact with conventional medication for diabetes. See also pp.42–51.

LEAVES

SOYA

Glycine max

Most familiar as soy sauce or tofu, soya appears to confer significant health benefits if used regularly. As food, soya is rich in protein, lecithin, and essential fatty acids; as medicine, it contains isoflavones and other compounds, which have oestrogenic and antioxidant activity.

MEDICINAL USES

Parts used Beans ● Sprouts

Key actions Oestrogenic ● Nutritive

Menopausal problems Several clinical trials have shown that concentrated soya extracts help to relieve menopausal problems such as hot flushes. These phytoestrogen-rich extracts are an option when considering natural alternatives for the menopause, but are best taken only where other approaches have failed to work. Soya sprouts, beans, and fermented products – as part of a balanced diet – provide a good input of nutrients and phytoestrogens, especially during the menopause. Sprouts are preferable to beans, since they are more nutritious, richer in phytoestrogens and, unlike beans, do not impair absorption of vitamins and minerals, notably iron.

Raised cholesterol levels Soya lecithin is a useful supplement, with research endorsing its ability to lower raised cholesterol levels.

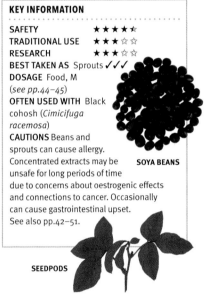

KEY INFORMATION

SAFETY	★ ★ ★ ★ ☆
TRADITIONAL USE	★ ★ ☆ ☆
RESEARCH	★ ★ ☆ ☆

BEST TAKEN AS Sprouts ✓✓✓
DOSAGE Food, M
(*see pp.44–45*)
OFTEN USED WITH Black cohosh (*Cimicifuga racemosa*)
CAUTIONS Beans and sprouts can cause allergy. Concentrated extracts may be unsafe for long periods of time due to concerns about oestrogenic effects and connections to cancer. Occasionally can cause gastrointestinal upset. See also pp.42–51.

SOYA BEANS

SEEDPODS

A staple food in much of Asia, soya has been used in China for at least 5000 years. It is today one of the world's most important food crops.

GINKGO

Ginkgo biloba

Harvested from the world's oldest surviving tree, ginkgo leaf extracts have been shown to improve blood flow to and through the arteries in the brain, to protect the central nervous system from oxidative damage, and to enhance mental recall and ability in healthy adults.

MEDICINAL USES

Parts used Leaf ● Fruit

Key actions Antioxidant ● Circulatory stimulant ● Improves mental performance ● Protects nerve tissue

Poor memory and recall Research evidence supports ginkgo's use to strengthen memory and cognitive function, though it is unclear just how far ginkgo alone is able to slow down the deterioration in mental function that occurs with dementia. Well-tolerated and rarely causing side effects, ginkgo is nonetheless a preferred initial treatment for these all too common conditions. Ginkgo can also be useful in treating the early stages of Parkinson's disease. It combines well with golden root (*Rhodiola rosea*).

Dizziness and tinnitus By improving blood flow to the central nervous system, ginkgo can benefit nerve-related problems such as dizziness, vertigo, nerve deafness, and tinnitus (ringing in the ears). Though such conditions can be very difficult to treat effectively, ginkgo is well worth trying for possible improvement in symptoms.

Leaves support healthy circulation to the hands and feet and can relieve restless legs.

Poor peripheral circulation Ginkgo stimulates blood flow throughout the body, from the head to the hands and feet. It can help with weak circulation, including altitude sickness, low blood pressure, Raynaud's syndrome, and intermittent claudication.

KEY INFORMATION

SAFETY	★ ★ ★ ⯪ ☆
TRADITIONAL USE	★ ★ ★ ★ ☆
RESEARCH	★ ★ ★ ★ ★

BEST TAKEN AS Standardized extract:
Tablet ✓✓✓ Tincture ✓✓
DOSAGE C *(see pp.44–45)*
CAUTIONS If taking prescribed medication, especially anticoagulant treatment, take only on the recommendation of a herbal or medical practitioner. Can occasionally cause headache or gastrointestinal upset. See also pp.42–51.

DRIED LEAVES

Other uses Given ginkgo's positive action on the circulation and central nervous system, it would be surprising were there no further indications for its use. In particular, ginkgo is taken on a daily basis by people aged 50 and over to support healthy circulation and mental function. As well as protecting nerve tissue from damage, it helps to ensure an adequate blood supply to the central nervous system. With professional advice, ginkgo can also be taken for conditions as varied as asthma, depression, frost bite, glaucoma, and multiple sclerosis.

Ginkgo trees are widely cultivated in China, France, and the USA.

BEYOND THE GREAT WALL

Once thought by western scientists to be extinct, ginkgoes were rediscovered in Japan in 1691. The trees had spread by seed from China, where they were mainly found in Buddhist monasteries, cultivated since c.1100 CE for medicinal uses.

PAGODA

LIQUORICE

Glycyrrhiza glabra

Tonic and anti-inflammatory, this most versatile of herbs finds use in treating ill health of all kinds. Added routinely to herbal prescriptions, liquorice acts on the adrenal glands and seems to reinforce the action and improve the flavour of herbs with which it is combined.

MEDICINAL USES

Part used Root

Key actions Anti-inflammatory ● Anti-viral ● Demulcent ● Expectorant ● Tonic

Gastritis, peptic ulcer, inflammatory bowel disease Liquorice's soothing, healing action works throughout the gastrointestinal tract, making it applicable in any situation where the gut or stomach wall is inflamed or ulcerated. Liquorice tea taken at night can help ease acid reflux.

Inflammatory arthritis Liquorice's anti-inflammatory action serves to relieve stiffness, heat, and pain in muscles and joints. Working in a manner not too dissimilar to prescribed steroids, it helps to dampen chronic inflammation, easing associated discomfort in conditions such as rheumatoid arthritis and polymyalgia rheumatica.

tincture is useful in many inflammatory conditions

TINCTURE

Mouth ulcers, sore throat, bronchitis, cough On its own or in combination with other herbs, liquorice tea makes an effective, pleasant-tasting mouthwash or gargle for sore tongue, mouth and throat ulcers, and laryngitis. Swallowed, its demulcent action soothes irritation and inflammation within the airways, such as in bronchial infection, helping to ease cough and stimulate the clearance of phlegm. The herb also appears to protect against dental caries.

Viral infections Not really a herb to use on its own in this context, liquorice combines well with other immune-enhancing herbs to strengthen the body's capacity to counter viral (and other) infections. Among other

FRESH ROOT

root has tonic properties

KEY INFORMATION

SAFETY	★★★★☆
TRADITIONAL USE	★★★★★
RESEARCH	★★★★☆

BEST TAKEN AS Tincture ✓✓✓ Capsule ✓✓ Dried root ✓
DOSAGE C *(see pages 44–45)*
OFTEN USED WITH Echinacea *(Echinacea purpurea)*
CAUTIONS Do not take large doses in high blood pressure. During pregnancy, or for long-term use, take only on the advice of a herbal or medical practitioner. See also pp.42–51.

DRIED ROOT PIECES

conditions, liquorice has been recommended for chronic fatigue syndrome, glandular fever, Lyme disease, shingles, and tonsillitis.

Adrenal tonic Liquorice can provide valuable support in any situation where the adrenal glands have been subject to long-term stress. It makes an effective tonic in aiding recovery from illness and chronic exhaustion. An oestrogenic remedy, it can be particularly helpful in menopausal exhaustion.

Other uses Research in China indicates that liquorice can prove helpful in polycystic ovary syndrome, improving menstrual regularity and fertility. It also supports liver function and soothes mucous membranes in the stomach and airways. Liquorice makes a valuable tonic to aid recovery from illness and the return to good health.

SWEET COURAGE

The medicinal value of liquorice was championed by the ancient Greek commander Alexander the Great (356–323 BCE). It is said his troops chewed on liquorice roots before a battle to give them fighting energy, which would have derived from its effect on their blood sugar and adrenal glands. Soldiers also used it to quench thirst while on the march, and thought it helped to stop them shaking with fright during battle.

ALEXANDER THE GREAT

Known as the "sweet herb", liquorice contains glycyrrhizic acid, which is 50 times sweeter than sugar.

A key astringent remedy for inflamed skin and eye disorders, witch hazel (*Hamamelis virginiana*) was used by early European settlers, who learnt of its properties from Native Americans.

WITCH HAZEL

Hamamelis virginiana

Produced by distilling the bark and leaves of this North American tree, witch hazel water is a useful skin cleanser and tonic. A frequent ingredient in cosmetics such as anti-wrinkle creams, the distilled water heals sore, roughened, or inflamed skin.

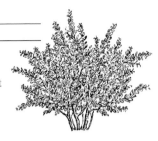

MEDICINAL USES

Parts used Bark ● Leaf

Key actions Anti-inflammatory ● Astringent ● Styptic (stops bleeding)

Eye problems For sore or inflamed eyes, grit, or an insect in the eye, soak cotton wool or lint in witch hazel water and place on the (closed) affected eye. Allow some liquid to enter the eye. For a better effect, add 5–10 drops of witch hazel water and a few grains of salt to an eyebath of clean warm water, and bathe the eye thoroughly, repeating as required.

Bruises, cuts, skin rashes, herpes sores Witch hazel is a useful cleanser for cuts and grazes, staunching blood flow at the same time. In weeping skin conditions, witch hazel water will help to dry up the leaking fluid and calm irritation. Mildly anti-viral, the water can lessen pain in herpes sores.

Varicose veins, capillary fragility, haemorrhoids Witch hazel tones and tightens irritated and over-relaxed tissue. Either witch hazel water or cream makes a first-rate application for varicose veins and haemorrhoids, controlling inflammation and toning distended veins. Thread veins can also benefit.

leaves and twigs are distilled to make witch hazel water

LEAVES

KEY INFORMATION

SAFETY ★ ★ ★ ★ ★
TRADITIONAL USE ★ ★ ★ ★ ☆
RESEARCH ★ ★ ★ ★ ☆
BEST TAKEN AS Topical: Distilled water or infusion ✓ ✓ ✓
DOSAGE T *(see pp.44–45)*
OFTEN USED WITH Aloe vera (*Aloe* spp.)
CAUTIONS None known. No longer used internally. See also pp.42–51.

FRESH BARK

A woodland tree native to Canada and eastern parts of the United States, witch hazel is today commonly grown for its attractive winter flowers.

DEVIL'S CLAW

Harpagophytum procumbens

With extensive traditional use for fevers, stomach ache, and rheumatic disease in its native southern Africa, devil's claw is now one of the most scientifically validated herbal medicines for rheumatic and arthritic disorders.

MEDICINAL USES

Part used Secondary root or tuber

Key actions Analgesic ● Anti-inflammatory ● Anti-rheumatic ● Bitter

Arthritic and rheumatic symptoms

Perhaps the first herbal remedy to consider for joint and muscle pain, devil's claw can relieve arthritic pain and inflammation, slowing or preventing deterioration in symptoms. Clinical studies have shown devil's claw extracts to be effective in relieving arthritic pain in the knee, hip, and back; patients were also able to reduce their intake of prescribed non-steroidal anti-inflammatories. Osteoarthritis, gout, fibromyalgia, and back pain can all benefit from this herb. Devil's claw is best taken before symptoms demand attention, especially if suitable dietary changes are also made. A distinctly bitter remedy, the tincture will help to stimulate appetite and absorption.

Devil's claw grows in semi-desert conditions and is increasingly rare in the wild. It is now being cultivated organically in Namibia; use organic where possible.

KEY INFORMATION

SAFETY ★ ★ ★ ☆ ☆
TRADITIONAL USE ★ ★ ★ ★ ⯨
RESEARCH ★ ★ ★ ★ ☆
BEST TAKEN AS Standardized extract: Tablet ✓ ✓ ✓
DOSAGE M (*see pp.44–45*)
OFTEN USED WITH Willow bark (*Salix alba*)
CAUTIONS With gallstones and peptic ulcer, take only on the advice of a herbal or medical practitioner. Do not take during pregnancy. May cause diarrhoea or interact with medications. See also pp.42–51.

tuber has anti-inflammatory properties

DRIED ROOT

CHOPPED ROOT

SEA BUCKTHORN

Hippophae rhamnoides

With high levels of minerals and vitamins A and C, sea buckthorn's tart-tasting berries make an ideal supplement to prevent colds and sore throats.

MEDICINAL USES

Parts used Fruit

Key actions Antioxidant • Circulatory tonic • Rich in vitamins and minerals

Infection As a natural supplement, the juice or syrup of sea buckthorn will improve resistance to colds, sore throat, and sinus problems.

Atherosclerosis, poor circulation to the retina and eye Rich in antioxidant bioflavonoids, sea buckthorn supports capillary and arterial health when taken long term.

BERRIES

KEY INFORMATION	
SAFETY	★ ★ ★ ★ ★
TRADITIONAL USE	★ ★ ★ ☆ ☆
RESEARCH	★ ★ ★ ☆ ☆
BEST TAKEN AS	Juice ✓✓✓
DOSAGE	M, C *(see pp.44–45)*
CAUTIONS	May increase the risk of bleeding, especially when combined with certain herbs or blood-thinning medications. See also pp.42–51.

HOPS

Humulus lupulus

Famed for their bitter taste when used to make beer, hops are a strong sedative and a common ingredient in many over-the-counter sleep remedies.

MEDICINAL USES

Parts used Flower (strobile)

Key actions Bitter tonic • Oestrogenic • Sedative

KEY INFORMATION	
SAFETY	★ ★ ★ ½ ☆
TRADITIONAL USE	★ ★ ★ ★ ☆
RESEARCH	★ ★ ★ ½ ☆
BEST TAKEN AS	Tincture ✓✓✓ Tablet ✓✓ Capsule ✓
DOSAGE	C *(see pp.44–45)*
CAUTIONS	Do not take in depression. Can cause drowsiness. See also pp.42–51.

Disturbed sleep To aid sleep, hops are best used in combination with other sedative remedies such as passion flower (*Passiflora incarnata*). Hops' oestrogenic action makes them potentially helpful in small doses for menopausal night sweats.

HOP STROBILES

Weak digestion, irritable bowel A few drops of hops tincture taken before meals stimulate appetite. With herbs such as chamomile (*Chamomilla recutita*), hops can help to relieve irritable bowel.

GOLDEN SEAL

Hydrastis canadensis

A potent herbal medicine, golden seal merits its reputation as a remedy that shifts chronic infection and heals weakened and congested mucous membranes. Its key use lies in the treatment of chronic bacterial, fungal, or viral infection affecting mucous membranes anywhere in the body.

MEDICINAL USES

Parts used Root

Key actions Anti-inflammatory ● Antimicrobial ● Antibacterial ● Blood cleanser ● Mucous membrane tonic ● Protects liver

Gastrointestinal infection, gastritis, and liver disorders Strongly bitter and detoxifying, golden seal exerts a positive influence on the stomach, intestines, and liver, helping in conditions as varied as peptic ulcer, dysbiosis, candidiasis, chronic gastroenteritis, and hepatitis.

Chronic infection Golden seal can significantly boost the body's ability to resist and shrug off lingering infection, be it a local fungal infection or glandular fever. It is often combined with echinacea (*Echinacea purpurea*).

Catarrhal problems Golden seal improves the health of mucous membranes. It is useful, for example, in sinus and middle ear congestion, particularly where linked to chronic infection, as well as in vaginal infection, where local application of a decoction may be helpful. For best results, combine with other herbs.

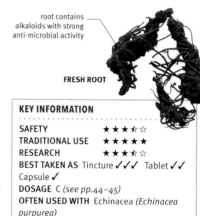

root contains alkaloids with strong anti-microbial activity

FRESH ROOT

Native to North American woodlands, golden seal is threatened in the wild through over-harvesting. Use cultivated, organic root only.

KEY INFORMATION

SAFETY ★ ★ ★ ⯪ ☆
TRADITIONAL USE ★ ★ ★ ★ ★
RESEARCH ★ ★ ★ ⯪ ☆
BEST TAKEN AS Tincture ✓✓✓ Tablet ✓✓ Capsule ✓
DOSAGE C *(see pp.44–45)*
OFTEN USED WITH Echinacea (*Echinacea purpurea*)
CAUTIONS Do not take during pregnancy and while breast-feeding. Not suitable for children. Keep to recommended dosage – high doses can irritate mucous membranes. See also pp.42–51.

ST JOHN'S WORT

Hypericum perforatum

Few herbs are as well-known as St John's wort, its popularity resulting from conclusive clinical research and first-hand experience of its effectiveness in treating mild to moderate depression. Other uses include seasonal affective disorder, jet lag, and nervous exhaustion.

MEDICINAL USES

Parts used Flowering top

Key actions Antidepressant • Anti-viral • Nerve tonic • Wound healer

Depression, disturbed sleep, seasonal affective disorder (SAD) Constituents of St John's wort influence brain chemistry in several different ways, leading to better mood and morale. Sleep, vitality, and the ability to relax may also improve. Positive results may take up to six weeks; more commonly, improvement starts within two weeks. The herb helps to banish the "winter blues", reducing the impact of SAD during the winter months. In European herbal medicine, St John's wort has always been seen as a remedy to drive away down-heartedness and to heal wounds, and these areas remain its core therapeutic uses. Clinical trials indicate that side effects from taking St John's wort are mild and very infrequent. Risks

KEY INFORMATION	
SAFETY	★ ★ ★ ★ ☆
TRADITIONAL USE	★ ★ ★ ★ ☆
RESEARCH	★ ★ ★ ★ ★
BEST TAKEN AS	Tincture ✓✓✓ Infusion ✓✓ Tablet, capsule, fixed oil (topically) ✓
DOSAGE	M, C *(see pp.44–45)*
OFTEN USED WITH	Valerian (*Valeriana officinalis*)
CAUTIONS	Can cause sensitivity to sunlight. If taking prescribed medication, including the contraceptive pill, seek advice from a herbal or medical practitioner before commencing treatment with St John's wort. Do not combine with other antidepressants. See also pp.42–51.

flowering tops are used to lift mood and morale

DRIED FLOWERS

in taking it arise only when it is used alongside certain conventional medicines since, by stimulating liver detoxification, St John's wort lowers drug levels within the body and thus reduces their effectiveness. In practice, this means that if taking prescribed medication, you should seek advice from a herbal practitioner or doctor before taking St John's wort.

Anxiety, nervous exhaustion Although emphasis is placed on the herb's ability to relieve depression, it also works well in relaxing and strengthening an exhausted nervous system such as can occur as a

St John's wort takes its name from the fact that the herb traditionally flowers by, and is harvested around, St John's Day on 24 June.

ST JOHN'S WORT OIL

To make the oil, collect the flowering tops of St John's wort on a dry, sunny morning. Carefully chop the herb material into 1cm (½in) lengths and place in a large clear glass jar. Pour on organic olive oil until the herb is fully covered. Stir the contents thoroughly and seal. Put on a sunny windowsill and leave for four weeks until the oil turns pale crimson. Strain and bottle, or for a stronger oil repeat, using a fresh batch of herb.

FIXED OIL

result of long-term stress or worry. It can be particularly beneficial in cases involving both anxiety and depression. St John's wort can help lift mood and vitality during menopause, while its effect on the hormone melatonin may make it useful for jet lag.

Toothache, sciatica, shingles St John's wort oil helps to dull nerve pain and speed tissue repair. Apply fixed oil directly on the skin overlying neuralgic areas or apply on the cheek before dental treatment.

Gastritis and stomach ulcer, local wound healing The herb's wound-healing properties can be put to good use, whether taken internally or applied externally. For stomach ulcers, take one teaspoonful of fixed oil a day. The fixed oil, which is naturally antiseptic, is traditionally regarded as a specific for knife and puncture wounds, and aids healing of wounds, sprains, bruises, and arthritic joints. The oil can also be applied to post-operative scars – a modern extension of its traditional use. Essential oils may be added to the oil for topical use, as required.

St John's wort has dark red oil glands that line the margins of both petals and leaves. These glands contain hypericin, a key active constituent.

ELECAMPANE

Inula helenium

Before the development of modern antibiotics, elecampane was a key remedy for treating tuberculosis. Today, it is more appropriate to think of it as a warming tonic and antibacterial remedy for less severe chest and throat infections, such as coughs, bronchitis, and tonsillitis.

MEDICINAL USES

Part used Root

Key actions Antibacterial ● Diaphoretic (induces sweating) ● Expectorant ● Mild bitter ● Tonic

Cough, chest infection, and congestion
Elecampane helps to relieve both dry irritable and wet catarrhal coughs. Best combined with other cough remedies, for example thyme (*Thymus vulgaris*), it may be safely given to children with chesty coughs. Its antibacterial action helps to disinfect the lungs, while congested phlegm is shifted and more readily coughed up. Elecampane combines well with remedies such as elderflower (*Sambucus nigra*) for catarrhal problems in the ear, nose, and throat, particularly where catarrh is dripping down the throat into the bronchial tubes. It can also prove useful in bronchial asthma.

KEY INFORMATION	
SAFETY	★ ★ ★ ★ ⯪
TRADITIONAL USE	★ ★ ★ ★ ☆
RESEARCH	★ ★ ★ ⯪ ☆

BEST TAKEN AS Tincture ✓✓✓ Decoction ✓✓ Tablet ✓ Capsule ✓
DOSAGE B *(see pp.44–45)*
OFTEN USED WITH Liquorice (*Glycyrrhiza glabra*)
CAUTIONS Rarely, can cause allergic reaction or gastrointestinal upset. Not advised in pregnancy and while breast-feeding. See also pp.42–51.

FRESH ROOT

Convalescence With a tonic action on both respiratory and digestive systems, elecampane makes a first-rate remedy for convalescence, improving appetite and restoring vitality, especially after a chest infection. The warm decoction, taken routinely throughout the winter, will help to protect against recurrence of infection.

Gastrointestinal parasites Elecampane has notable activity against intestinal worms and parasites. In combination with other remedies, it has been used successfully to treat amoebic dysentery.

A vigorous perennial growing up to 3m (10ft) in height, elecampane sends down deep roots that are harvested and dried in the autumn.

SHIITAKE

Lentinus edodes

Reputedly given to the Japanese emperor in 199 CE, shiitake mushroom has been used in traditional medicine for centuries as a tonic and restorative remedy, often for severe and chronic illness. Recent research suggests that shiitake extract has promise as an anti-cancer remedy.

MEDICINAL USES

Part used Fungi

Key actions Anti-tumour ● Anti-viral ● Immune-enhancing ● Protects liver

used to make "medicinal" soup

DRIED MUSHROOMS

Anti-cancer remedy Following extensive Japanese research, fairly high doses of shiitake extract are recommended in cancer treatment, lower dose long-term treatment being used to support immune function and act as a cancer preventative. As an anti-cancer agent, shiitake appears to be most effective when taken alongside chemotherapy, the extract acting to stimulate programmed cell death and to protect the liver and immune system from toxic damage. In this situation, shiitake extract should be taken only on the advice of a suitably experienced herbal practitioner or doctor.

Other uses Many chronic health problems such as viral hepatitis, chronic fatigue syndrome, candidiasis, and recurrent infections can benefit from shiitake's positive action on immune function. Shiitake can also be taken to maintain good health, especially in old age. For medicinal use, take shiitake extract. Shiitake mushrooms make a good addition to a healthy diet.

Shiitake grows naturally on fallen broadleaf trees in temperate regions of eastern Asia, including China and Japan. It is now widely cultivated in the West.

KEY INFORMATION

SAFETY	★ ★ ★ ★ ☆
TRADITIONAL USE	★ ★ ★ ★ ★
RESEARCH	★ ★ ★ ⯪ ☆

BEST TAKEN AS Standardized extract, fungi (as food) ✓✓✓
DOSAGE M, Food (*see pp.44–45*)
OFTEN USED WITH Reishi mushroom (*Ganoderma lucida*)
CAUTIONS Can cause diarrhoea and allergic skin reactions, as well as elevation of blood counts. See also pp.42–51.

LAVENDER

Lavandula angustifolia, L. officinalis, L. vera

Lavender combines beauty and function, delicate aroma with great therapeutic use, and an exceptional safety record. Its dried flowers and essential oil revive the spirits and at the same time earn a place in every home's first aid box. Little wonder that lavender is so popular.

MEDICINAL USES

Parts used Essential oil ● Flower

Key actions Analgesic ● Antidepressant ● Antispasmodic ● Relieves wind ● Sedative ● Antiseptic

Anxiety, irritability, headache Anxiety and tension are key words for lavender, its calming qualities soothing nervous overactivity, stress-related headache, and migraine. It may perhaps also reduce the risk of fits and convulsions. Lavender combines well with rosemary (*Rosmarinus officinalis*) to alleviate nervous exhaustion and improve weak circulation. It is also known to have mood-enhancing properties. This combination of mild sedative and antidepressant activity makes it particularly suitable where lowered mood and vitality follow long-term worry or overactivity.

Sleep difficulties Whether used as essential oil (a few drops placed on a burner or used in a massage), dried flowers (in a herb pillow), or as a tincture (a teaspoon before going to bed), lavender improves sleep quality and the chances of a good night's sleep.

ESSENTIAL OIL

DRIED FLOWERS

> **KEY INFORMATION**
>
> | | |
> |---|---|
> | SAFETY | ★ ★ ★ ★ ★ |
> | TRADITIONAL USE | ★ ★ ★ ★ ☆ |
> | RESEARCH | ★ ★ ★ ½ ☆ |
>
> **BEST TAKEN AS** Essential oil (topically) ✓✓✓ Tincture ✓✓ Infusion ✓
> **DOSAGE** B (see pp.44–45)
> **OFTEN USED WITH** Rosemary (*Rosmarinus officinalis*)
> **CAUTIONS** Rarely, when applied topically, the oil can cause contact dermatitis. Do not take the essential oil internally. See also pp.42–51.

It combines well with other herbal sleep remedies, notably passion flower (*Passiflora incarnata*) and lemon balm (*Melissa officinalis*).

Local pain relief and relaxation

Lavender oil can be applied to the skin in almost any situation involving pain. Massage the oil onto herpes or shingles sores, rheumatic joints, the cheek overlying an aching tooth, or the forehead and temples for migraine. A few drops of oil on cotton wool plugged into the ear relieves mild earache. Lavender tincture or tea can provide systemic support, helping to relax tensed and aching muscles. For menstrual cramps and pain, tincture or tea taken internally and oil massaged into the lower abdomen can bring quick relief.

Digestive problems Most often combined with other digestive remedies, lavender has specific value where emotional distress is the underlying cause of digestive disturbance such as "butterflies" in the stomach, bloating, belching, and irritable bowel symptoms.

Other uses In France, lavender is used in the treatment of a wide range of respiratory problems, including flu, cough, whooping cough, asthma, and bronchitis. Its antispasmodic and antiseptic activity combines well with other herbs such as thyme (*Thymus vulgaris*) and elecampane (*Inula helenium*). The flowers are sometimes used as a gargle for painful and inflamed sore throat; in this situation, lavender combines well with liquorice (*Glycyrrhiza glabra*). Lavender tea or tincture has long-standing use as a tonic for weak and nervy children; neither taste as good as they smell, and they can be diluted in unsweetened blackcurrant or apple juice.

Native to France and the western Mediterranean, lavender is cultivated worldwide for its volatile oil.

LAVENDER - THE ESSENTIAL OIL

Possibly the only essential oil that is safe to apply neat to large areas of the skin, lavender oil is a key first aid remedy for home use and when travelling. A few drops massaged into the temples or the back of the neck can bring swift relief to headache and neck and shoulder tension. Aches and pains elsewhere in the body will also benefit. Apply the oil to minor burns, sunburn, sores, itchy skin problems, and insect bites and stings, to promote healing and ease discomfort. The oil also makes a reasonably effective insect repellent.

PAIN RELIEF

Aromatic and relaxing, lavender (*Lavandula officinalis*) is cultivated widely for its medicinal properties. Its healing and pain-relieving essential oil is a favourite first aid remedy.

MOTHERWORT

Leonurus cardiaca

An undervalued remedy, motherwort has an unusual combination of actions, all tending to calm those with a nervous disposition, while strengthening cardiovascular and digestive function. The herb's botanical name indicates its long-standing use as a remedy for an irregular or fast heartbeat.

MEDICINAL USES

Part used Flowering top

Key actions Aids menstruation ● Heart tonic ● Lowers blood pressure ● Mild bitter ● Nerve tonic ● Relaxant

Menstrual problems Used to promote a regular and symptom-free menstrual cycle, motherwort relieves pre-menstrual tension and period pains. It stimulates the onset of menstruation, and is also

DRIED AERIAL PARTS

helpful where an irregular or absent menstrual cycle is linked to poor appetite or low body weight.

Heart and thyroid problems
Motherwort is a key remedy for palpitations and irregular heartbeat, especially when they are linked with anxiety or an overactive thyroid. Small, frequent doses (a few drops of tincture) can sometimes be sufficient to control such problems. It is prescribed by herbal practitioners for angina, coronary heart disease, and high blood pressure.

Used for centuries to treat heart palpitations, motherwort, a perennial herb, grows in much of Europe and North America.

KEY INFORMATION

SAFETY	★ ★ ★ ★ ★
TRADITIONAL USE	★ ★ ★ ★ ☆
RESEARCH	★ ★ ☆ ☆ ☆

BEST TAKEN AS Infusion ✓✓✓
Tincture ✓✓ Capsule ✓
DOSAGE C *(see pp.44–45)*
OFTEN USED WITH Cramp bark (*Viburnum opulus*)
CAUTIONS Do not take during pregnancy. Avoid in heavy menstrual bleeding. See also pp.42–51.

LEMON VERBENA

Lippia citriodora syn. *Aloysia triphylla*

Native to South America, lemon verbena is best used as a refreshing tea. It has sedative and relaxant properties, and can be used as an insect repellent.

MEDICINAL USES

Part used Leaf

Key actions Antispasmodic ● Sedative

Sleep difficulties The gentle-acting infusion of lemon verbena leaves makes a pleasant after-dinner and late-evening drink, aiding relaxation and helping to prevent insomnia and restlessness. The herb has a mild tonic effect on the nervous system, which lifts the spirits and helps counter depression.

Digestive aid The lemon-scented volatile oil in the infusion improves digestion and soothes discomfort in the stomach, calming wind and bloating.

FRESH LEAVES

> **KEY INFORMATION**
>
> SAFETY ★★★★☆
> TRADITIONAL USE ★★★★☆
> RESEARCH ★★☆☆☆
> BEST TAKEN AS Infusion ✓✓✓
> DOSAGE C (*see pp.44–45*)
> CAUTIONS None known. See also
> pp.42–51.

LOBELIA

Lobelia inflata

A valuable remedy in the treatment of respiratory problems, lobelia is potentially toxic and should be taken only when prescribed or as a licensed medicine.

MEDICINAL USES

Parts used Aerial parts

> **KEY INFORMATION**
>
> SAFETY ★★☆☆☆
> TRADITIONAL USE ★★★★★
> RESEARCH ★★☆☆☆
> BEST TAKEN AS Over-the-counter
> remedy ✓✓✓
> DOSAGE M (*see pp.44–45*)
> CAUTIONS May cause vomiting in high
> dosage. Take only when prescribed by a
> herbal or medical practitioner, or when part
> of a licensed over-the-counter medicine.
> Restricted herb in some countries, including
> Australia. See also pp.42–51.

Key actions Antispasmodic ● Expectorant ● Stimulates breathing

Chest complaints Lobelia relaxes the airways, stimulates coughing up of mucus, and eases wheezing in the chest. It promotes deeper and stronger breathing, and is a valuable remedy for tight-chested conditions such as asthma and chronic bronchitis.

FLOWERS AND LEAVES

LINSEED, FLAX

Linum usitatissimum

Grown as a food crop in temperate climates, linseed, or flax, is a valuable and readily available dietary supplement. A rich source of protein and omega-3 oils, linseed also contains high levels of phytoestrogens – roughly 10 times more than other seeds, making it a key remedy for the menopause.

MEDICINAL USES

Part used Seed

Key actions Antioxidant ● Demulcent ● Laxative ● Nutritive ● Oestrogenic

Food supplement, menopausal symptoms Ground or cracked seeds (untreated seeds are not absorbed) make an excellent addition to the diet: take 1–2 tablespoons a day with muesli, breakfast cereal, or yoghurt. As the seeds soak up large quantities of liquid, drink a large glass of water at the same time. The alpha-linolenic acid and omega-3 oil content in the seeds is similar to fish oils, though less biologically available. High levels of phytoestrogens make linseed a useful supplement for menopausal symptoms such as hot flushes and headache. Store ground or cracked seeds in an airtight container in the refrigerator to prevent the seed oils from going rancid. Use within two to three weeks.

Digestive problems An excellent bulk laxative, linseed is a safe and frequently effective remedy for chronic constipation. Soak 1 tablespoonful of seeds in at least 5 times its volume of warm water. Leave

SEEDS seeds are an effective remedy for constipation

for a few hours, then swallow, preferably drinking additional water. The resulting jelly-like brew will often prove helpful for constipation and can also relieve acid indigestion and diarrhoea. Long-term problems such as acid reflux and oesophagitis, peptic ulcer, and chronic constipation are likely to need ongoing treatment with linseed.

CRUSHED SEEDS crushed seeds are a rich source of omega-3 fatty acids

Respiratory disorders Taken in the form of soaked seed as described for digestive problems (*see above*), linseed soothes the chest and airways and can prove helpful in conditions such as chronic or irritable coughs, hoarseness, bronchitis, pleurisy, and emphysema.

KEY INFORMATION	
SAFETY	★ ★ ★ ★ ☆
TRADITIONAL USE	★ ★ ★ ★ ☆
RESEARCH	★ ★ ★ ★ ☆
BEST TAKEN AS	Ground or cracked seed ✓✓✓
DOSAGE	1–2 tablespoons a day, with plenty of water.
CAUTIONS	Always dilute with at least 5 times the volume of water. Unripe seed can be toxic. See also pp.42–51.

The seeds may also be applied to the chest wall as a poultice (*see Topical use, below*) to relieve congestive bronchitis.

Topical use Linseed also finds use when applied locally to the skin as a poultice. Put warm soaked seeds in muslin and hold or bind in place on burns, bites and stings, boils, and haemorrhoids. A warm poultice can also be used to "draw" splinters and boils, the mucilage in the seeds soaking up fluids and waste products.

Other uses Although research is not so far conclusive, there are indications that linseed can prove a useful remedy in a remarkably wide range of serious health problems. It appears to have a soothing effect on the kidneys and may be helpful in kidney disease. Within the gut it helps to prevent absorption of fats and sugars, and makes a good addition in both cholesterol-lowering regimens and diet-controlled diabetes. Partly as a result of its high phytoestrogen content, linseed appears to have

Cultivated for food in Africa since 5000–8000 BCE, linseed was first brought to North America for its stem fibre to make linen and paper.

important cancer-preventative activity, particularly against oestrogen-dependent cancers such as breast cancer. It is thought to reduce re-absorption of oestrogens within the colon, and is likely to prove useful in the treatment of colon cancer. In the above conditions, use linseed only on the advice of your doctor or herbal practitioner.

LINSEED OR FLAX OIL

Flax or linseed oil contains uniquely high levels (typically around 55 per cent) of alpha-linolenic acid, an omega-3 polyunsaturated fatty acid similar to those found in fish oils. Plant-derived omega-3 oils are not so readily available for use by the body as fish oils, but nevertheless provide similar health benefits.

All omega-3 oils have a protective activity on the heart and circulation and against cancer.

ALPHA-LINOLENIC ACID MOLECULE

With long-standing medicinal as well as industrial use, linseed (*Linum usitatissimum*) is better known today as the main vegetable source of omega-3 essential fatty acids.

ALFALFA, LUCERNE

Medicago sativa

Most used as animal feed, alfalfa is equally nutritious for humans, containing appreciable levels of protein, calcium, magnesium, vitamins C, E, K, and beta-carotene. Its main traditional use has been as a natural food supplement for debility and convalescence, aiding appetite and weight gain.

MEDICINAL USES

Parts used Aerial parts • Sprouts

Key actions Appetite stimulant • Oestrogenic • Nutritive • Stimulates breast milk

Convalescence and debility Alfalfa infusion and sprouts provide high-quality and easily absorbed nutrition, especially when taken medium term. Indications for their use include poor appetite, convalescence, inability to gain weight, and anorexia nervosa.

Menopausal symptoms Alfalfa is a useful food to supplement during the menopause. Unlike soya, it does not inhibit the absorption of minerals such as iron and calcium. Alfalfa contains oestrogenic isoflavones, which has led to its recent use for menopausal

seeds contain oestrogenic isoflavones

SEEDS

symptoms,especially in combination with sage (*Salvia officinalis*).

Other uses Alfalfa is also useful for arthritic and rheumatic symptoms, diabetes, and raised cholesterol levels.

With easily assimilated nutrients, alfalfa's combination of phytoestrogens, calcium, and magnesium can prove useful in the prevention of osteoporosis.

KEY INFORMATION	
SAFETY	★ ★ ★ ★ ☆
TRADITIONAL USE	★ ★ ★ ⯪ ☆
RESEARCH	★ ★ ★ ☆ ☆
BEST TAKEN AS	Sprouts ✓✓
	Leaf infusion ✓✓✓
DOSAGE	M, A, food *(see pp.44–45)*
OFTEN USED WITH	Sage (*Salvia officinalis*)
CAUTIONS	Keep to recommended dosage. Do not eat excessive amounts of sprouting seeds. Avoid during pregnancy. See also pp.42–51.

TEA TREE

Melaleuca alternifolia

In aboriginal medicine tea tree leaves, crushed and inhaled or infused, were employed to treat infections of all kinds. Today, the essential oil is normally used, its unrivalled antiseptic action proving effective at countering fungal infection affecting the hair, skin, and nails.

MEDICINAL USES

Parts used Essential oil ● Leaf

Key actions Antifungal ● Antiseptic

Skin infections A must have for any home herbal first aid kit, tea tree oil can help treat many minor fungal and bacterial skin problems. On boils, acne spots, and small patches of fungal infection, for example, an affected toenail, apply neat tea tree oil sparingly twice a day. For larger areas, apply tea tree oil diluted in calendula (*Calendula officinalis*) or wheatgerm (*Triticum vulgare*) oil. Combine 1 part tea tree oil to 20 parts carrier oil and massage onto the affected area.

Ear infection For infection of the external ear passage and for mild earache, put 1–2 drops of neat oil on cotton wool and plug into the affected ear overnight. Tea tree oil combines well with lavender oil – use 1 drop of each.

Vaginal infection For vaginal infections such as thrush, diluted tea tree oil can be applied to the affected area though it is likely to sting. It is far better to use tea tree pessaries. Insert one every night for 3–4 days, stop for a few days, and then start again as required.

KEY INFORMATION

SAFETY ★ ★ ★ ☆ ☆
TRADITIONAL USE ★ ★ ★ ★ ★
RESEARCH ★ ★ ★ ★ ☆
BEST TAKEN AS Neat or diluted essential oil (topical) ✓✓✓ Pessary ✓✓
DOSAGE T (*see pp.44–45*)
OFTEN USED WITH Calendula fixed oil
CAUTIONS Do not take essential oil internally. Occasionally can cause dermatitis with topical preparations. See also pp.42–51.

Native to Australia, tea tree flourishes in moist soils in northern New South Wales and Queensland. The leaves are distilled to produce the essential oil.

leaves contain highly antiseptic volatile oil

FRESH LEAVES

LEMON BALM, MELISSA

Melissa officinalis

A much-loved remedy, lemon balm's soothing qualities quieten the heart and an overactive mind. Valuable in easing anxiety and mild depression, research indicates that the herb helps to improve cognition, strengthen memory and mental focus, and has actions that counter the development of Alzheimer's disease.

MEDICINAL USES

Part used Leaf

Key actions Antidepressant
- Antispasmodic • Insect repellent
- Relaxant • Relieves wind
- Topical (anti-viral)

Anxiety, tension headache, insomnia, palpitations Lemon balm is a relaxing tonic for anxiety, mild depression, restlessness, and insomnia. It reduces feelings of panic and is a valuable remedy for palpitations of a nervous origin. For all such conditions, take as infusion or tincture 2–3 times a day.

leaves are used as a nerve tonic

DRIED LEAVES

Memory and mental focus Even short term use of lemon balm is thought to help improve brain function and strengthen memory and mental performance. A clinical trial found that people with mild to moderate Alzheimer's disease had improved cognition and reduced agitation when taking lemon balm.

Other uses An infusion or tincture, dabbed on a cold sore or insect bite can bring quick relief. Safe for children, lemon balm is a specific for stress-related stomach disorders such as acidity, indigestion, bloating, and colic.

With an aromatic and pleasant bitter-lemon taste, lemon balm is a common ingredient in liqueurs and *digestifs,* including Benedictine.

KEY INFORMATION

SAFETY	★ ★ ★ ★ ★
TRADITIONAL USE	★ ★ ★ ★ ☆
RESEARCH	★ ★ ★ ⯪ ☆

BEST TAKEN AS Infusion (fresh leaves) ✓✓✓ Tincture ✓✓
DOSAGE C *(see pp.44–45)*
OFTEN USED WITH Rosemary (*Rosmarinus officinalis*)
CAUTIONS None known. See also pp.42–51.

PEPPERMINT

Mentha x piperita

A mint hybrid first grown in England in the 17th century, peppermint is known the world over for its cool and flavoursome taste. Commonly taken after a meal to aid digestion, the infusion is a useful remedy for wind, flatulence, and bloating, as well as headaches and migraines linked to digestive weakness.

MEDICINAL USES

Parts used Leaf ● Essential oil

Key actions Antiseptic ● Antispasmodic ● Diaphoretic ● Mild analgesic ● Mild bitter ● Mild sedative ● Relieves wind

Windy digestion, nausea, cramps, irritable bowel Clinical research confirms the usefulness of peppermint essential oil in irritable bowel syndrome. The essential oil acts on the colon, relieving spasm and irritability, and reducing the sensitivity of nerve endings in the intestinal wall. The milder-acting infusion can be safely taken for symptoms such as bad breath, wind, belching, bloating, and colic.

Colds, flu, headache, migraine Peppermint and elderflower (*Sambucus nigra*) make an effective combination for fever, colds, catarrh, and gastric

KEY INFORMATION

SAFETY	★ ★ ★ ★ ☆
TRADITIONAL USE	★ ★ ★ ★ ☆
RESEARCH	★ ★ ★ ★ ☆

BEST TAKEN AS Infusion ✓✓✓ Tincture ✓✓ Enteric-coated capsules (on prescription only) ✓
DOSAGE C (*see pp.44–45*)
OFTEN USED WITH Elderflower (*Sambucus nigra*)
CAUTIONS Do not give to children under the age of 5. Essential oil is best taken internally only on the recommendation of a herbal or medical practitioner.
 See also pp.42–51.

infection. Hot peppermint tea encourages sweating and cools fever. Drink tea or apply 1–2 drops of peppermint oil to forehead to relieve headache and migraine.

INFUSION

Topical uses Peppermint oil is soothing for itchy skin. Apply at 2 per cent dilution (2 drops per teaspoonful of carrier oil) to affected areas. The infusion can be applied as a lotion to relieve nettle rash and eczema. Be careful to avoid the eyes.

DRIED AERIAL PARTS

Peppermint leaf makes a good after-dinner drink.

BOGBEAN

Menyanthes trifoliata

Strongly bitter, bogbean is mostly used to improve a weak or underactive digestive system or to treat rheumatic symptoms, whether resulting from local or systemic inflammation. It is a threatened species because its natural habitat is disappearing; use organic products.

MEDICINAL USES

Part used Leaf

Key actions Anti-rheumatic ● Bitter tonic ● Laxative (large doses)

Arthritic, rheumatic, and kidney problems Generally prescribed by herbal practitioners rather than self-medicated, bogbean can provide relief in conditions as varied as fibromyalgia, gout, polymyalgia rheumatica, and rheumatoid arthritis.

Seen as a specific for muscular, rather than joint, aches and pains, it combines well with other anti-rheumatic herbs, including birch (*Betula alba*) and celery (*Apium graveolens*). It has a reputation of supporting kidney clearance of waste products and has been used in kidney disease.

Loss of appetite, weak digestion Bogbean stimulates appetite and the flow of saliva and digestive juices, leading to better processing of foods and absorption of nutrients. An irritant laxative at high doses, it should be avoided where the bowels are loose or sensitive such as in irritable bowel.

leaves have a strong bitter taste

DRIED LEAVES

Unglamorously but accurately named, bogbean thrives in boggy, marshy places and fresh shallow water.

KEY INFORMATION

SAFETY ★ ★ ★ ★ ☆
TRADITIONAL USE ★ ★ ★ ⯪ ☆
RESEARCH ★ ★ ⯪ ☆ ☆
BEST TAKEN AS Tincture ✓✓✓
DOSAGE C *(see pp.44–45)*
OFTEN USED WITH Celery seed (*Apium graveolens*)
CAUTIONS May cause diarrhoea. Avoid during pregnancy and while breast-feeding. See also pp.42–51.

BASIL

Ocimum basilicum

The herb in Italian cooking, basil relieves upper digestive discomfort, indigestion, and bloating, and is applied topically to acne spots and insect bites.

MEDICINAL USES

Part used Whole plant

Key actions Antibacterial ● Insecticidal ● Mild sedative ● Relieves wind

Digestive problems Like many culinary herbs, basil has a pronounced effect on the digestion, stimulating and at the same time soothing stomach and intestinal activity. It is best taken for symptoms such as bad breath, stomach cramps, nausea, indigestion, wind, and bloating.

Other uses Juice from the leaves can be applied neat to infected spots and insect bites and stings to speed healing.

Significantly insecticidal, the leaves or juice can be rubbed on the skin to repel insects.

AERIAL PARTS

KEY INFORMATION

SAFETY	★ ★ ★ ★ ☆
TRADITIONAL USE	★ ★ ★ ☆ ☆
RESEARCH	★ ★ ★ ☆ ☆

BEST TAKEN AS Infusion ✓✓✓ Capsule ✓✓ Food ✓
DOSAGE C (*see pp.44–45*)
CAUTIONS Safe at recommended dosage. See also pp.42–51.

HOLY BASIL, TULSI

Ocimum sanctum

Prized in Indian medicine as a tonic which clears the mind, holy basil has many benefits, including stabilizing blood sugar levels and soothing chest conditions.

MEDICINAL USES

Part used Whole plant

Key actions Anti-inflammatory ● Expectorant ● Lowers blood sugar levels ● Tonic

Uses Holy basil promotes better uptake of sugars within the body, and can prove particularly helpful in the early stages of diabetes. Commonly used for cough and bronchitis, holy basil also helps to lower blood pressure and cholesterol.

FRESH LEAVES

KEY INFORMATION

SAFETY	★ ★ ★ ★ ☆
TRADITIONAL USE	★ ★ ★ ★ ★
RESEARCH	★ ★ ★ ★ ☆

BEST TAKEN AS Infusion ✓✓✓ Tincture ✓✓ Powder ✓
DOSAGE C (*see pp.44–45*)
CAUTIONS In diabetes, take only on professional advice. See also pp.42–51.

A prolific flowering herb, evening primrose (*Oenothera biennis*) produces thousands of seeds, several hundred of which need to be pressed for just one capsule of seed oil.

EVENING PRIMROSE

Oenothera biennis

Researched in England since the 1980s, evening primrose seed oil (EPO) is high in omega-6 essential fatty acids and can prove helpful in a range of inflammatory conditions, such as menstrual problems, rheumatoid arthritis, and eczema. It is thought to work in two distinct ways to block inflammatory processes.

MEDICINAL USES

Parts used Seed ● Seed oil

Key actions Anti-inflammatory ● Antioxidant ● Emollient

Skin conditions Less concentrated than borage oil but with similar therapeutic effects, EPO makes a useful supplement for eczema and dermatitis. It is considered safe for infants as well. For best results, take internally for several months and apply topically to affected areas of skin.

Menstrual disorders Studies show that EPO can bring relief to pre-menstrual syndrome (PMS), especially where breast or menstrual pain are key problems. In both cases, EPO is best combined with a vitamin B complex supplement.

Inflammation EPO can act to reduce inflammation in chronic inflammatory disease, easing symptoms in joints and muscles, such as in rheumatoid arthritis.

KEY INFORMATION

SAFETY	★ ★ ★ ★ ★
TRADITIONAL USE	★ ★ ★ ☆ ☆
RESEARCH	★ ★ ★ ★ ☆

BEST TAKEN AS Capsule ✓✓✓ Oil (topically) ✓✓
DOSAGE M *(see pp.44–45)*
OFTEN USED WITH Vitamin E
CAUTIONS Safe at recommended dosage. Avoid in pregnancy. Take only the advice of a herbal or medical practitioner if taking epilepsy medication. See also pp.42–51.

SEED-OIL CAPSULE

EPO can also help ease the discomfort of dry eyes and deficient skin secretion in Sjögren's syndrome.

Grown commercially for its seed oil, evening primrose is a native of North America and thrives on wasteland, especially in sandy soil.

OREGANO, WILD MARJORAM

Origanum vulgare, O. majorana

Common to cuisines of the Mediterranean, the aromatic, slightly spicy flavour of oregano adds zest to food, while stimulating digestive activity. The herb is used for digestive disorders and throat or chest infections. Essential oils from oregano species have strong antiseptic and antifungal activity.

MEDICINAL USES

Parts used Aerial parts ● Essential oil

Key actions Antioxidant ● Antiseptic ● Antifungal ● Expectorant ● Stimulant

Respiratory and digestive infection
With strongly antiseptic and antimicrobial constituents, oregano infusion or tincture is a useful expectorant in bronchial infection, chesty coughs, and respiratory catarrh. Digestive problems such as gastroenteritis and candida infection will also benefit from the herb's tonic activity, especially where bloating and food intolerance are present. For mouth and throat infections, including oral thrush, use the infusion as a mouthwash or gargle, then swallow.

Other uses Apply the infusion or the diluted oil (a maximum of 5 per cent dilution in a carrier oil such as olive oil) regularly to skin problems such as ringworm and fungal nails.

An upright perennial with aromatic, oval leaves, oregano or cultivated marjoram is a common kitchen garden herb, used often in Mediterranean cooking.

wild plant is weaker in action than garden varieties

DRIED LEAVES AND FLOWERS

WILD MARJORAM

KEY INFORMATION

SAFETY   ★ ★ ★ ★ ☆
TRADITIONAL USE   ★ ★ ★ ★ ☆
RESEARCH   ★ ✫ ☆ ☆ ☆
BEST TAKEN AS Infusion ✓✓✓
Tincture ✓✓ Capsule ✓
DOSAGE C (*see pp.44–45*)
OFTEN USED WITH Olive leaf
(*Olea europaea*)
CAUTIONS Do not take during pregnancy. Do not take essential oil internally unless recommended by a herbal or medical practitioner. Can cause skin irritation and gastrointestinal upset. See also pp.42–51.

WHITE PEONY

Paeonia lactiflora

A traditional Chinese remedy taken to cool excess heat, white peony is mostly used in the West for period pains, menstrual irregularity, and fibroids.

MEDICINAL USES

Part used Root

Key actions Anti-inflammatory
● Antispasmodic ● Tonic

Gynaecological problems Although it is taken most often to ease period pains and to treat fibroids, when combined with liquorice (*Glycyrrhiza glabra*) white peony acts to reverse the symptoms of polycystic ovary syndrome and may help to improve fertility in this condition.

FLOWER

KEY INFORMATION

SAFETY	★ ★ ★ ★ ☆
TRADITIONAL USE	★ ★ ★ ★ ☆
RESEARCH	★ ★ ★ ☆ ☆

BEST TAKEN AS Decoction ✓✓✓
Tincture ✓✓
DOSAGE B (*see pp.44–45*)
CAUTIONS Do not take if taking prescribed medicines to thin blood. Avoid during pregnancy. Rarely, can cause gastrointestinal upset. See also pp.42–51.

GUARANA

Paullinia cupana

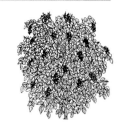

With up to 7 per cent caffeine, guarana is a popular Brazilian drink taken to boost energy and alertness. It is used by athletes to enhance peak performance.

MEDICINAL USES

Part used Seed (roasted)

Key actions Antioxidant ● Astringent
● Stimulant

Low vitality and stamina Guarana is taken much like coffee to ward off fatigue, maintain attention and alertness, and to support maximal energy output, for example, in sports competition. Like other caffeine-containing products, it can prove helpful in relieving tension headache and migraine. It has been used symptomatically to treat diarrhoea. However, it is often too stimulant for those with chronic health problems.

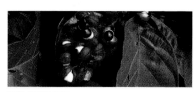

FRUIT WITH SEEDS

KEY INFORMATION

SAFETY	★ ★ ★ ★ ☆
TRADITIONAL USE	★ ★ ★ ★ ☆
RESEARCH	★ ★ ★ ☆ ☆

BEST TAKEN AS Tablet ✓✓✓ Drink ✓✓
DOSAGE M (*see pp.44–45*)
CAUTIONS Avoid excessive doses. Avoid if pregnant or breast-feeding. See also pp.42–51.

PASSIFLORA, PASSION FLOWER

Passiflora incarnata

Used long before the arrival of Europeans in the Americas, passiflora's calming, relaxant qualities are quickly apparent on taking the herb, and underlie its many uses. Passiflora is most often employed to relieve anxiety and nervousness and to aid sleep, its use in this respect confirmed by clinical trials.

MEDICINAL USES

Part used Whole plant

Key actions Aids sleep ● Relaxant ● Relieves pain ● Sedative

Anxiety, nervousness, racing heart, headache Rarely producing drowsiness, passiflora takes the edge off worry and anxiety, bringing relief to symptoms such as a racing heart and tension headache. A mild analgesic, passiflora can help with migraine and neuralgic pain, for example, in toothache.

Sleep disturbance, spasmodic pain Safe and non-addictive, passiflora is a key sleep remedy, often enabling one to relax and slip off into sound sleep. Its relaxant and antispasmodic activity, which is commonly overlooked, also finds use in conditions such as leg cramps and period pains.

DRIED AERIAL PARTS | aerial parts are used to make relaxing infusions

Native to southern USA, Central and South America, passiflora is today cultivated extensively in Europe.

tincture is analgesic and antispasmodic

TINCTURE

KEY INFORMATION

SAFETY	★ ★ ★ ★ ★
TRADITIONAL USE	★ ★ ★ ★ ⯪
RESEARCH	★ ★ ★ ⯪ ☆

BEST TAKEN AS Tincture ✓✓✓ Tablet ✓✓
DOSAGE C (*see pp.44–45*)
OFTEN USED WITH Valerian (*Valeriana officinalis*)
CAUTIONS Rarely, can cause allergic reactions. See also pp.42–51.

A key sedative remedy, passion flower (*Passiflora incarnata*) was first brought to Europe from Brazil by Jesuit priests. The complex structure of the flower was taken to signify Christ's passion.

GINSENG

Panax ginseng

Ginseng has a mystique of its own, its documented use in the Far East extending back to pre-history. Taken over the centuries by kings, emperors, and popes, it has an unrivalled reputation for improving overall vitality and acting as a male tonic.

MEDICINAL USES

Part used Root

Key actions Adaptogen
● Anti-inflammatory ● Antioxidant
● Immune tonic ● Tonic

Chronic ill health, fatigue, convalescence Traditional Chinese use emphasizes ginseng's restorative qualities, helping to strengthen in depleted states and promoting longevity. Taken through the long winter months of northern China by the frail and elderly, ginseng aids physical endurance, particularly in those with lowered vitality and poor immune function. Low doses taken long term are most likely to prove effective.

Short-term enhancement of mental and physical performance At recommended dosage, ginseng will help to increase muscle weight, physical strength and stamina, and improve mental ability. Standard advice for healthy adults is to take for a maximum of 6 weeks. A general recommendation is to avoid caffeine while taking ginseng.

Ginseng is now rarely found in the wild. It is widely cultivated in China and Korea using intensive farming methods.

Male tonic Ginseng is probably best thought of as a male tonic rather than as an aphrodisiac. It can certainly help to improve physiological and sexual function and may also increase sexual vitality. It is a herb of choice in treating erectile dysfunction, where it combines well with ginkgo (*Ginkgo biloba*) or golden root (*Rhodiola rosea*), and for low sperm count.

TINCTURE

Clinical trials and quality control
Ginseng has been the subject of intensive research, with clinical trials investigating a wide range of potential therapeutic applications for the herb. Clinical evidence supports ginseng's use in the following ways: to enhance physical and mental performance, including coping with hunger and extremes of temperature; to improve sperm count (in men with low sperm count); to reduce

KEY INFORMATION

SAFETY	★ ★ ★ ★ ☆
TRADITIONAL USE	★ ★ ★ ★ ★
RESEARCH	★ ★ ★ ★ ☆

BEST TAKEN AS Tincture or tablet: standardized to 4% ginsenosides ✓✓✓
DOSAGE 0.5–2g a day or M (see pp.44–45)
OFTEN USED WITH Ginkgo (*Gingko biloba*)
CAUTIONS Do not take during acute illness. In high blood pressure and diabetes, take only on the advice of a herbal or medical practitioner. See also pp.42–51.

THE KING OF HERBS

Regularly taken by the emperors of China and their households, ginseng's first documented use is in a Chinese herbal written 2,000 years ago: "Ginseng is a tonic to the five viscera, quieting the animal spirits, stabilizing the soul, preventing fear, expelling the vicious energies, brightening the eye, improving vision... and prolonging life."
Shen'nong Bencaojing (Shen Nong's Materia Medica), 1st century CE.

CHINESE EMPEROR

Other uses Ginseng is most likely to prove helpful in treating chronic infection and a depleted immune system when used in combination with other immune-modulating remedies such as astragalus (*Astragalus membranaceus*) and echinacea (*Echinacea* spp.). Ginseng helps to enhance memory and recall, and combined with ginkgo (*Ginkgo biloba*), can be useful for the elderly as a preventative for dementia. With cancer-preventative activity, ginseng can be a helpful restorative when recovering from cancer (seek professional advice here). The root appears to help control blood sugar levels in diabetes, and it can relieve chronic or recurrent headache when linked to overwork or exhaustion.

FRESH ROOT

menopausal fatigue and tiredness; to improve immune function and resistance to infection; and perhaps most importantly, to improve quality of life. Root and root extracts need to be of good quality for positive results and it is important to select a reputable brand or supplier; ideally, choose a standardized tincture or extract. Given the relatively high cost of ginseng root, adulterated products are not uncommon.

The root is harvested after four years; the older the root, the better its strength and quality.

ginseng extracts help to improve stamina

TABLETS

BUTTERBUR

Petasites hybridus

Formerly used to treat plague, butterbur has long been used in Europe as a cough and cold remedy, and for stomach and gall bladder problems. Clinical trials suggest butterbur extract provides effective relief for migraine and hay fever, and the extract is now commonly available over the counter.

MEDICINAL USES

Part used Root

Key actions Anti-allergenic
● Anti-inflammatory ● Antispasmodic

Migraine, pain relief In clinical trials, butterbur extract reduced the frequency, duration, and intensity of migraine attacks. A safe treatment for migraine, the extract can be given to children from the age of six, but only at the advised dosage. The extract may also help to relieve joint and menstrual pain.

Allergic rhinitis With its marked anti-allergenic activity, butterbur can bring allergic rhinitis, such as hay fever, under control. Butterbur can alleviate its troublesome symptoms such as nasal congestion, sore throat, and sneezing, especially if combined with an appropriate diet.

Coughs, colds, chest infections Although not the first cough and cold remedy that comes to mind, butterbur extract can prove useful in speeding up recovery from respiratory problems such as chesty coughs and bronchitis.

Gastrointestinal disorders Research suggests that butterbur reduces ulceration of the small intestine.

KEY INFORMATION

SAFETY	★ ★ ★ ★ ☆
TRADITIONAL USE	★ ★ ★ ½ ☆
RESEARCH	★ ★ ★ ★ ☆

BEST TAKEN AS Take *only* as a standardized extract ✓ ✓ ✓
DOSAGE M (*see pp.44–45*)
CAUTIONS Do not take in pregnancy or while breast-feeding. Take only purified products where liver-toxic compounds have been removed. Rarely, may cause gastrointestinal upset or drowsiness. Restricted herb in some countries; not legally available in Australia. See also pp.42–51.

The wild plant, but not the extract, is toxic to the liver.

DRIED ROOT

PARSLEY

Petroselinum crispum

A useful food at any time, parsley is rich in a number of readily absorbable nutrients, including vitamin C and phytoestrogens, making it a valuable supplement, particularly during the menopause. Medicinally, the root is preferred, having a distinct benefit on the urinary tract and in rheumatic problems.

MEDICINAL USES

Parts used Leaf ● Root

Key actions Anti-rheumatic
● Antispasmodic ● Diuretic
● Nutritive ● Relieves wind
● Stimulates menstruation

Menopausal symptoms, prevention of osteoporosis Moderately oestrogenic, parsley leaf is a nutritious food supplement to take during the menopause. Its relatively high boron content makes it a valuable supplement in natural approaches to preventing osteoporosis.

Urinary tract problems Commonly used with other urinary antiseptic remedies, parsley root can bring relief to the urinary tract in disorders such as mild cystitis and urethritis. It has traditionally been used in the prevention and treatment of kidney stones, and is thought to aid the kidneys in the clearance of waste products that exacerbate muscle aches and stiffness.

Other uses Parsley has strong deodorizing properties and the leaf is commonly chewed to treat bad breath

Parsley leaf is both flavoursome and rich in nutrients, notably significant levels of vitamins C and E, iron, boron, and phytoestrogens.

and to freshen the breath. It is said to mask the odour of garlic on the breath. The root has a tonic activity on the digestion, helping to relieve indigestion, wind, and bloating. Valued for its ability to promote menstrual blood flow, parsley at the recommended dosage can help in stimulating regular menstruation and relieving menstrual cramps. Parsley is thought to suppress breast milk production, so it is best avoided when breast-feeding. The root can be used to relieve arthritic symptoms.

DRIED LEAVES

KEY INFORMATION

SAFETY	★ ★ ★ ☆ ☆
TRADITIONAL USE	★ ★ ★ ★ ★
RESEARCH	★ ★ ★ ★ ☆
BEST TAKEN AS	Tincture ✓✓✓
DOSAGE	C (*see pp.44–45*)
OFTEN USED WITH	Meadowsweet (*Filipendula ulmaria*)

CAUTIONS Do not use medicinally during pregnancy or while breast-feeding. See also pp.42–51.

BLACK CATNIP, PHYLLANTHUS

Phyllanthus amarus

A central Ayurvedic remedy for the liver, research into black catnip has produced conflicting results. Evidence suggests that it has liver-protective activity.

MEDICINAL USES

Parts used Whole plant

Key actions Anti-viral ● Diuretic ● Liver protector ● Lowers blood sugar levels

Liver disease Traditional use indicates that black catnip can help in disorders such as viral hepatitis, gall bladder disease, and gallstones. A safe remedy, it is best taken on professional advice.

plant has diuretic qualities

LEAF

Viral infections Following the herb's traditional use, black catnip makes a useful herb to combine with immune-modulating remedies such as echinacea (*Echinacea* spp.) in treating viral and other chronic infections.

KEY INFORMATION

SAFETY	★ ★ ★ ★ ☆
TRADITIONAL USE	★ ★ ★ ★ ✬
RESEARCH	★ ★ ★ ☆ ☆
BEST TAKEN AS	Decoction ✓✓✓ Tincture ✓✓ Tablet ✓
DOSAGE	M, B *(see pp.44–45)*
CAUTIONS	Rarely, may cause mild side effects. See also pp.42–51.

POMEGRANATE

Punica granatum

Recent research has focused on pomegranate juice as a preventative for prostate and colon cancer, and as a support for a healthy heart and circulation.

MEDICINAL USES

Part used Fruit

Key actions Antioxidant ● Anti-viral ● Supports heart and circulation

FRESH FRUIT

KEY INFORMATION

SAFETY	★ ★ ★ ★ ★
TRADITIONAL USE	★ ★ ★ ☆ ☆
RESEARCH	★ ★ ★ ✬ ☆
BEST TAKEN AS	Juice ✓✓✓
DOSAGE	As food; M *(see pp.44–45)*
CAUTIONS	Use fruit (seeds) or juice only; take rind only if professionally prescribed.

Cancer preventative and cardio-vascular support With its very high levels of polyphenols, pomegranate juice clearly has powerful antioxidant activity. This makes the juice a valuable supplement to help prevent cancer development. The juice lowers blood pressure and supports heart and arterial health. A small glassful a day is a good tonic for the cardio-vascular system.

Other uses A traditional use of pomegranate juice (blended with seeds) is in the treatment of diarrhoea.

PLANTAIN

Plantago major, P. lanceolata

Known in Gaelic as "the healing herb", plantain is a versatile herb that will benefit many conditions. Mostly used to support and strengthen mucous membranes throughout the body, plantain counters infection, reduces mucous membrane secretions, and supports tissue repair.

MEDICINAL USES

Parts used Whole plant

Key actions Analgesic ● Anticatarrhal ● Antihaemorrhagic ● Anti-inflammatory ● Anti-viral ● Demulcent ● Wound healer

Catarrhal problems All catarrhal problems from ears to chest, and throughout the digestive tract, will benefit from plantain. It may be taken for colds, hay fever, sinusitis, cough, and sore throat. It also treats acid indigestion, peptic ulcer, diarrhoea, and irritable bowel. Best taken as tea, the tincture will also serve well. Take short-term for acute conditions, but long-term for chronic states such as sinus congestion, allergic rhinitis, and mucus colitis.

Broken or inflamed tissue An effective external application wherever tissue repair is needed, plantain also stimulates healing within the body, notably in the digestive and respiratory tracts.

ribwort plantain

FLOWERSPIKE   **DRIED LEAVES**

Greater plantain (*P. major*) is preferred medicinally (*right*), but ribwort plantain (*P. lanceolata*) can be used in its place (*top left*).

KEY INFORMATION

SAFETY	★ ★ ★ ★ ☆
TRADITIONAL USE	★ ★ ★ ★ ☆
RESEARCH	★ ★ ★ ☆ ☆

BEST TAKEN AS Infusion ✓✓✓
Tincture ✓✓
DOSAGE B *(see pp.44–45)*
OFTEN USED WITH Elderflower (*Sambucus nigra*)
CAUTIONS Rarely, may cause dermatitis.
See also pp.42–51.

Cultivated since antiquity, the rose (*Rosa* spp.) is known as the "queen of flowers" for its beauty and fragrance. Rosewater distilled from the flowers is tonic to the skin.

REHMANNIA

Rehmannia glutinosa

A tonic for the liver and kidneys, rehmannia is also an important anti-inflammatory herb. In its native China, it is used to cool fever in chronic illnesses.

MEDICINAL USES

Part used Root

Key actions Adrenal tonic ● Anti-haemorrhagic ● Anti-inflammatory ● Reduces fever

Chronic inflammation A key herb used to control inflammation in chronic inflammatory disorders such as rheumatoid arthritis and polymyalgia rheumatica (especially where exhaustion and weakness are factors), rehmannia is best taken on professional advice.

FLOWERS AND LEAVES

KEY INFORMATION

SAFETY ★ ★ ★ ★ ☆
TRADITIONAL USE ★ ★ ★ ★ ☆
RESEARCH ★ ★ ★ ☆ ☆
BEST TAKEN AS Tincture ✓✓✓
Decoction ✓✓
DOSAGE B *(see pp.44–45)*
CAUTIONS In pregnancy, take only on professional advice. Occasionally can cause diarrhoea. See also pp.42–51.

ROSE

Rosa spp.

Rosewater, distilled from the petals, soothes and tones the skin. A syrup from the hips of the dog rose (*R. canina*) is given to children to improve resistance to infection.

MEDICINAL USES

Part used Petals ● Fruit (hips)

Key actions Anti-inflammatory ● Astringent ● Tonic

KEY INFORMATION

SAFETY ★ ★ ★ ★ ★
TRADITIONAL USE ★ ★ ★ ★ ★
RESEARCH ★ ★ ☆ ☆ ☆
BEST TAKEN AS Syrup ✓✓✓
Aromatic water (topically) ✓✓✓
DOSAGE C *(see pp.44–45)*
CAUTIONS Rarely, fruit or syrup may cause gastrointestinal upset. See also pp.42–51.

Skin toner Rosewater makes a valued application for lax tissue, including burns, mouth ulcers, chapped hands, and sore eyes.

FLOWER

CHINESE RHUBARB

Rheum officinale

A well-tolerated and moderately powerful laxative, Chinese rhubarb exerts a positive influence on the liver, gall bladder, and intestines. At low dose, its astringent actions predominate to control diarrhoea, while at normal dose, it acts as a mild irritant laxative.

MEDICINAL USES

Part used Root

Key actions Antibacterial ● Astringent ● Bitter ● Blood cleanser ● Laxative

Constipation Chinese rhubarb root is best used as a short-term remedy for constipation when other approaches, such as increased fibre in the diet or linseed, fail. Take two 0.5g capsules with chamomile or ginger tea in the evening. Repeat for up to two weeks. If the problem persists, seek professional advice.

Other uses In combination with other remedies, Chinese rhubarb has significant therapeutic use in gastrointestinal infection and chronic inflammation such as Crohn's disease.

Chinese rhubarb root has mild antimicrobial activity against a range of bacteria, fungi, and viruses.

KEY INFORMATION

SAFETY	★ ★ ★ ★ ☆
TRADITIONAL USE	★ ★ ★ ★ ☆
RESEARCH	★ ★ ★ ★ ☆

BEST TAKEN AS Capsule ✓✓✓ Tincture ✓✓
DOSAGE C (*see pp.44–45*)
OFTEN USED WITH Ginger (*Zingiber officinale*)
CAUTIONS Do not take during pregnancy or while breast-feeding. Do not give to children. May cause gastrointestinal upset. See also pp.42–51.

root is an effective laxative

POWDERED ROOT

GOLDEN ROOT, ARCTIC ROOT

Rhodiola rosea

Found in mountainous regions and tundra as far north as the Arctic, golden root has benefits similar in many ways to ginseng. A key remedy for long-term stress and physical and mental fatigue, it supports the body's stress response.

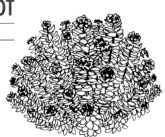

MEDICINAL USES

Part used Root

Key actions Adaptogen
● Anti-inflammatory
● Antioxidant ● Protects
heart ● Tonic for males

Physical endurance and long-term stress An important endurance herb, golden root helps the body and mind to adapt efficiently to increased physical and mental demands, as in sports training or studying for exams. It also improves work performance.
It can be valuable in depleted states such as chronic fatigue and nervous exhaustion, though care should be taken to start at a low dose and increase it gradually.

FRESH ROOT

root extract relieves stress

Golden root can be taken to prevent altitude sickness.

KEY INFORMATION

SAFETY	★ ★ ★ ★ ☆
TRADITIONAL USE	★ ★ ★ ★ ☆
RESEARCH	★ ★ ★ ★ ☆

BEST TAKEN AS Standardized extract (rosavin): Tablet, tincture ✓✓✓
DOSAGE M *(see pp.44–45)*
OFTEN USED WITH Ginseng (*Panax ginseng*)
CAUTIONS Can cause irritability and sleep disturbance. Not advisable in manic and bipolar disorders. See also pp.42–51.

TABLETS

Depressive mood Though golden root does not as yet have established anti-depressant activity, it does appear to help raise mood and vitality in those who have a tendency to suffer from depression.

BLACKCURRANT

Ribes nigrum

The tart, crisp flavour of blackcurrant reflects the fruit's high vitamin C content and marked antioxidant activity. Long given to children to protect them from colds and throat infection, the fruit, juice, and extracts are thought to have significant anti-inflammatory and antiviral activity.

MEDICINAL USES

Parts used Fruit ● Bud ● Leaf ● Seed oil

Key actions Adrenal support (buds) ● Anti-inflammatory ● Antioxidant ● Astringent ● Diuretic (leaves)

Immune support Following traditional use, blackcurrant juice or extract can be taken regularly to maintain healthy resistance to viral infection, including colds, flu and herpes sores. It is likely to speed recovery from infection as well. Avoid sweetened concentrates.

Allergy and chronic inflammation In France, the buds are thought to support adrenal gland function and are prescribed to treat allergic and inflammatory conditions such as asthma. The seed oil, which possesses high levels of omega-6 essential fatty acids (GLA) with similar properties to evening primrose oil, can be taken

Blackcurrant is grown mainly in eastern Europe for its sour-sweet fruit. The leaves and berries are harvested in early and late summer respectively.

leaves are used as a diuretic

FRESH LEAF

to control inflammation in conditions such as rheumatoid arthritis and multiple sclerosis.

Circulatory tonic With their high levels of antioxidant anthocyanins, the fruit and leaves can be taken long-term as a treatment to strengthen the circulation, including the capillaries. The juice or extract can be taken daily for several months to help improve circulatory problems such as capillary fragility and varicose veins.

KEY INFORMATION

SAFETY	★ ★ ★ ★ ★
TRADITIONAL USE	★ ★ ★ ★ ☆
RESEARCH	★ ★ ★ ☆ ☆
BEST TAKEN AS	Juice/extract ✓✓✓

Infusion ✓✓ Seed oil ✓
DOSAGE Food; M (*see pp.44–45*)
CAUTIONS None known. See also pp.42–51.

juice is high in vitamin C and antioxidants

BLACKCURRANT JUICE

A Mediterranean herb, rosemary (*Rosmarinus officinalis*) symbolizes fidelity between lovers, an association that may have been made on account of its ability to improve memory.

ROSEMARY

Rosmarinus officinalis

Few herbs are as well known as rosemary, especially for its distinctive aroma. Traditionally used to strengthen memory and recall, it is frequently taken to aid study and exam performance, and to ward off mental exhaustion.

MEDICINAL USES

Parts used Leaf ● Essential oil

Key actions Anti-inflammatory ● Antioxidant ● Antispasmodic ● Circulatory stimulant ● Digestive tonic ● Nerve tonic

Headache, migraine, nervous exhaustion Rosemary can bring quick relief to headaches caused by overwork and nervous tension. For headaches linked to high blood pressure, combine with limeflowers (*Tilia* spp.). It can also prove helpful in migraine.

Digestion, poor circulation, low energy Tonic and antioxidant, rosemary stimulates digestion and blood flow throughout the body, proving helpful for those with low energy levels, especially where linked to low blood pressure or poor appetite. It is a key herb for those failing to thrive, either after long-term illness or where digestion and circulation are weak. For best results, take rosemary tea or tincture before meals for several months.

Hair tonic An infusion made from the leaves acts as a natural hair conditioner, toning the scalp and strengthening the hair.

KEY INFORMATION	
SAFETY	★ ★ ★ ★ ☆
TRADITIONAL USE	★ ★ ★ ★ ⯨
RESEARCH	★ ★ ★ ☆ ☆
BEST TAKEN AS Infusion ✓✓✓ Tincture ✓✓ Essential oil (topically) at maximum 5% dilution in carrier oil ✓	
DOSAGE C (*see pp.44–45*)	
OFTEN USED WITH Lavender (*Lavandula officinalis*)	
CAUTIONS Rarely, may cause contact allergy. See also pp.42–51.	

leaves are used to relieve headaches and tension

DRIED LEAVES

Rosemary tea or diluted essential oil makes an excellent rub for sore and aching muscles and joints.

RASPBERRY

Rubus idaeus

Known in classical times as an aid to childbirth, raspberry leaf is thought to act on the womb, relaxing the cervix and toning the muscles that contract during labour. Recent studies indicate that it is a safe remedy that shortens labour and reduces the likelihood of a forceps delivery.

MEDICINAL USES

Part used Leaf

Key actions Aids preparation for childbirth ● Anti-diarrhoeal ● Astringent

To aid preparation for childbirth
Take raspberry leaf as an infusion or capsule on a daily basis for the last three months of pregnancy, and freely during labour to ease contractions. The normal daily dose is 1–2 cups of tea.

FRESH LEAF

DRIED LEAVES

leaves make an effective gargle for sore throat

Heavy menstrual bleeding Raspberry leaf can help to control heavy menstrual bleeding, combining well with yarrow (*Achillea millefolium*) to reduce blood loss. It is sometimes taken to relieve pre-menstrual symptoms and period pains.

Other uses With its strong astringent action, an infusion of the leaves makes a gentle-acting and effective remedy for diarrhoea and loose bowel movements in children. It is equally useful as a gargle for sore throats and a lotion for sore eyes.

Raspberry is grown mostly for its delicious-tasting fruit, which can be made into a syrup or vinegar; traditionally, it was used to treat feverish states.

KEY INFORMATION

SAFETY	★ ★ ★ ☆ ☆
TRADITIONAL USE	★ ★ ★ ★ ★
RESEARCH	★ ★ ★ ★ ☆

BEST TAKEN AS Dried flower bud ✓✓✓ Essential oil ✓✓ Tincture ✓
DOSAGE C (*see pp.44–45*)
OFTEN USED WITH Liquorice (*Glycyrrhiza glabra*)
CAUTIONS Do not take during the first 3 months of pregnancy. During the last 3 months, best taken on the advice of a herbal or medical practitioner. See also pp.42–51

A venous tonic, butcher's broom (*Ruscus aculeatus*) has leaf-like leathery branches with a terminal spine. Traditionally, it was used as a diuretic and to treat kidney disorders.

SHEEP'S SORREL

Rumex acetosella

A small dock-type plant, sheep's sorrel is rarely used as a medicine on its own. Its chief therapeutic use lies in its role as the principal remedy in the Essiac formula.

MEDICINAL USES

Part used Whole plant

Key actions Anti-inflammatory ● Diuretic ● Laxative

Claimed cancer cure The Essiac formula (a cancer treatment devised in the 1920s by the Canadian nurse Rene Caisse, following a native Ojibwa recipe) comprises sheep's sorrel, burdock (*Arctium lappa*), slippery elm (*Ulmus rubra*), and Chinese rhubarb (*Rheum officinale*). No detailed investigation has yet taken place into its clinical effects. Burdock and Chinese rhubarb, but not sheep's sorrel, are known to possess anti-cancer activity, and prepared

Essiac tea has strong antioxidant properties.

AERIAL PARTS

> **KEY INFORMATION**
>
> SAFETY ★★☆☆☆
> TRADITIONAL USE ★★☆☆☆
> RESEARCH ★☆☆☆☆
> BEST TAKEN AS Infusion ✓✓✓
> Tincture ✓✓
> DOSAGE C *(see pp.44–45)*
> CAUTIONS Do not take excess doses of root. Do not give to children. Take only on professional advice. Occasionally may cause gastrointestinal upset or skin reactions. See also pp.42–51.

BUTCHER'S BROOM

Ruscus aculeatus

An unusual-looking native European plant with stiff spiky "leaves", butcher's broom has been investigated in detail as a remedy for problems affecting the veins.

MEDICINAL USES

Part used Root

Key actions Anti-inflammatory ● Laxative ● Venous tonic

Venous insufficiency Butcher's broom contracts vein walls, leading to reduced fluid loss into surrounding areas. It may need to be taken long-term for varicosed leg veins and to reverse fluid retention in the lower legs. Unlike horse chestnut (*Aesculus hippocastanum*), it has little positive effect on the arteries.

> **KEY INFORMATION**
>
> SAFETY ★★★★½
> TRADITIONAL USE ★★★½☆
> RESEARCH ★★★★☆
> BEST TAKEN AS Capsule ✓✓✓
> Tablet ✓✓✓ Tincture ✓
> DOSAGE M *(see pp.44–45)*
> CAUTIONS Occasionally may cause gastrointestinal upset. See also pp.42–51.

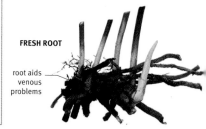

FRESH ROOT

root aids venous problems

YELLOW DOCK

Rumex crispus

A common weed and wayside plant, yellow dock has a deep taproot that draws up iron and other minerals from the soil, presenting them in an accessible form for absorption. The root is therefore prescribed in cases of iron-deficiency anaemia, although its main use is for sluggish bowels and mild constipation.

MEDICINAL USES

Parts used Root ● Leaf (topically)

Key actions Detoxifier ● Laxative

Chronic skin problems Yellow dock is best used in combination with other herbs rather than on its own. It fits well in formulae that contain other "blood cleansers" such as burdock (*Arctium lappa*) to support bowel clearance and liver detoxification. Yellow dock is called for in conditions involving chronic toxicity, including skin disorders such as acne and boils, eczema, and psoriasis. Often the best approach is to take small amounts regularly over several months, promoting gradual but effective detoxification. Other types of chronic illness which involve poor elimination can also benefit from the root's cleansing activity, for example swollen glands and throat infection.

Other uses Yellow dock combines well with nettle (*Urtica dioica*) in allergic and rheumatic conditions. Both herbs also contain appreciable levels of iron and can be used as a natural iron supplement in mild cases of anaemia.

KEY INFORMATION

SAFETY	★ ★ ★ ★ ☆
TRADITIONAL USE	★ ★ ★ ★ ☆
RESEARCH	★ ☆ ☆ ☆ ☆

BEST TAKEN AS Tincture ✓✓✓
Decoction ✓✓
DOSAGE C (*see pp.44–45*)
OFTEN USED WITH Burdock (*Arctium lappa*)
CAUTIONS Do not take excess doses of root. Do not take leaves internally. Not advisable during pregnancy. See also pp.42–51.

Yellow dock is a traditional remedy for nettle stings; rub the fresh leaves firmly onto the affected area. Do not take the leaves internally as they are poisonous.

root is useful for skin complaints

DRIED ROOT

WILLOW BARK, WHITE WILLOW

Salix alba

An ancient remedy for aches and pains, fevers, and rheumatic conditions, willow bark contains aspirin-like substances. It is often thought of as the herbal equivalent of aspirin, but its mode of action is only partly the same – it cannot be used as a straightforward aspirin replacement.

MEDICINAL USES

bark relieves stiffness

FRESH BARK

Part used Bark

Key actions Analgesic ● Anti-inflammatory ● Astringent ● Relieves fever

Aches and pains The bark may be taken as a first aid remedy for headache, toothache, and back pain. Its main use is in muscle and joint inflammation, pain and stiffness, and for conditions such as sports injuries and gout. The herb causes few side effects and may be preferable to aspirin-type anti-inflammatories in conditions such as osteoarthritis that require long-term use.

Fever Take an infusion (perhaps with ginger, *Zingiber officinale*) to control fevers and to relieve the malaise and discomfort that accompanies acute infection. If the temperature is 39°C (102.2°F) or above, seek professional advice without delay.

KEY INFORMATION

SAFETY   ★ ★ ★ ★ ☆
TRADITIONAL USE   ★ ★ ★ ★ ★
RESEARCH   ★ ★ ★ ☆ ☆
BEST TAKEN AS Tincture ✓✓✓ Capsule ✓✓ Infusion ✓
DOSAGE A (*see pp.44–45*)
OFTEN USED WITH Devil's claw (*Harpagophytum procumbens*)
CAUTIONS Do not take if allergic to aspirin, or while breast-feeding. Do not give to children with viral infections. Can cause allergic reactions. May occasionally cause gastrointestinal upset. See also pp.42–51.

Native to much of Europe, willow bark thrives in damp areas and on river banks. It is thought to treat "damp" conditions within the body.

DAN SHEN, ASIAN RED SAGE

Salvia miltiorrhiza

A cousin of garden sage, dan shen is *the* Chinese
remedy for heart and circulation, and has been used
for over 2,000 years. The root has an impressive
range of activity on the cardiovascular system,
benefiting conditions such as high blood pressure,
poor peripheral circulation, and a failing heart.

MEDICINAL USES

Part used Root

Key actions Anticoagulant ● Heart
and circulatory tonic ● Lowers blood
pressure ● Sedative

Heart tonic Similar in some respects
to hawthorn (*Crataegus* spp.), dan shen
increases blood flow through the
coronary arteries and has a marked
relaxant action on the circulation to
the heart, making it a key remedy
for angina and a weak or under-
performing heart. Dan shen is best
taken on the recommendation of a
herbal or medical practitioner.

**High blood pressure and poor
peripheral circulation** Dan shen relaxes
arterial circulation and thins the blood,
both factors contributing to a lower
blood pressure and stronger circulation
to the hands and feet.

Other uses The herb has strong
anticoagulant activity – seek professional
advice if taking blood-thinning
medication or where a tendency
to bleeding or bruising exists.

A native Chinese herb, dan shen is cultivated
in northeastern China and Inner Mongolia, and
sold in herbal markets across China as a
circulatory stimulant.

KEY INFORMATION

SAFETY	★ ★ ★ ☆ ☆
TRADITIONAL USE	★ ★ ★ ★ ★
RESEARCH	★ ★ ★ ★ ☆

BEST TAKEN AS Decoction ✓✓✓
Tincture ✓✓ Tablet ✓
DOSAGE C (*see pp.44–45*)
CAUTIONS Do not take during
pregnancy. Do not take with prescribed
blood-thinning medication. May
occasionally cause gastrointestinal
upset. See also pp.42–51.

DRIED CHOPPED ROOT

root has strong
anticoagulant
activity

tincture is taken
to relieve angina

TINCTURE

The bluish-purple flowers of sage (*Salvia officinalis*) make a sharp contrast to its grey-green leaves. The Latin name *Salvia* means "to cure", echoing the medieval perception of the herb.

CHIA SEED

Salvia hispanica

Chia seeds have been used as a foodstuff to make tortillas, tamales, and drinks in traditional Mexican and Central American cooking since pre-Columbian times. Harvested from a native sage plant, they contain very high levels of oil. They have become increasingly popular in North America and Europe over the last 10 years and are now a common food supplement and ingredient in natural health foods.

MEDICINAL USES

SEEDS

Part used Seed

Key actions Nutritive
● Demulcent

Nutritional food Chia seeds make a highly nutritious food, comparable in many ways to linseed (*Linum usitatissimum*). Both contain high levels of oil and protein, and a fibre-rich seed coating that becomes jelly-like on being soaked in water. Chia oil contains roughly 60 per cent alpha-linolenic acid, an omega-3 oil, and 20 per cent linoleic acid, an omega-6 oil. Though these plant-sourced oils are less biologically available than those found in fish oil, chia still makes a good dietary source of these key essential fatty acids. To date, there is little indication that chia has a pronounced positive effect on blood sugar and cholesterol levels.

Other uses Like many other seeds, notably linseed, chia contains appreciable levels of phytoestrogens, and can be taken as part of a diet to help rebalance oestrogen levels at the menopause. When soaked in water, chia seeds produce a thick jelly-like brew that has a demulcent or mucus-like action on the body, and can be helpful in relieving acid indigestion and constipation.

Chia seed is thought to contain higher levels of omega-3 oils than any other plant or seed.

> **KEY INFORMATION**
>
> SAFETY ★★★★★
> TRADITIONAL USE ★★★★★
> RESEARCH ★☆☆☆☆
> BEST TAKEN AS Seed ✓✓✓
> DOSAGE ½–1 tbsp a day
> OFTEN USED WITH Linseed (*Linum usitatissimum*)
> CAUTIONS None known

SAGE, SPANISH SAGE

Salvia officinalis, S. lavandulifolia

In 1551, the English herbalist William Turner noted that sage "restores natural heat ... comforts the vital spirits ... helps the memory, and quickens the senses". Such praise is not misplaced for sage, which is truly a versatile and effective tonic.

MEDICINAL USES

Part used Leaf

Key actions Antimicrobial
● Antioxidant ● Astringent
● Digestive tonic ● Oestrogenic
● General tonic ● Reduces sweating

Mouth ulcers, sore throat, excess catarrh With its valuable astringent properties, sage counters infections, such as mouth ulcers and sore throat, and dries up catarrh. The infusion makes an excellent mouthwash and gargle for local infections. Sage combines well with herbs such as echinacea (*Echinacea* spp.) for recurrent problems.

Menopausal symptoms Cold sage tea sipped in small doses through the day is a traditional and frequently effective remedy for menopausal symptoms such as sweating, hot flushes, and headache.

Poor memory, stress, and anxiety Recent research points to sage as a potential remedy for early stages of dementia. Its tranquillizing properties help relieve stress and anxiety, and also improve mental vitality and memory. Spanish sage is preferable for long-term use – its low thujone content makes it safer than ordinary sage.

Despite its medicinal value, sage is most commonly known as a culinary herb.

TINCTURE

KEY INFORMATION

SAFETY ★ ★ ★ ✬ ☆
TRADITIONAL USE ★ ★ ★ ★ ✬
RESEARCH ★ ★ ★ ☆ ☆
BEST TAKEN AS Infusion ✓✓✓ Tincture ✓✓ Capsule ✓
DOSAGE C *(see pp.44–45)*
OFTEN USED WITH Alfalfa (*Medicago sativa*)
CAUTIONS Do not take during pregnancy and while breast-feeding. Excessive doses can be toxic. See also pp.42–51.

DRIED LEAVES

SCHISANDRA, WU WEI ZI

Schisandra chinensis

Known in China as "the five-flavoured herb" for the distinctively sour and slightly salty flavour of its berries, schisandra has been traditionally used as a sexual tonic for both men and women. The herb is thought to tone the kidneys and sexual organs, protect the liver, and improve mental stamina.

MEDICINAL USES

Part used Fruit

Key actions Adaptogen ● Antioxidant ● Mild antidepressant ● Protects liver ● Tonic

Reduced vitality A few schisandra berries chewed every day will increase physical and mental vitality and help combat stress. Schisandra is often combined with ginkgo (*Ginkgo biloba*) to boost memory and improve concentration. With its mild sedative and antidepressant qualities, schisandra can help with depressive states linked to long-term stress and mental exhaustion. Schisandra can add zest to life and for this reason can be a valuable tonic where libido is low.

Liver disorders Schisandra improves liver health and aids in the effective metabolism of toxins. Chronic liver disorders in general can benefit, including chronic viral hepatitis. In this situation, take the herb only on professional recommendation.

Other uses Schisandra is used in the treatment of respiratory infections such as chronic cough, shortness of breath,

It is said that if taken for 100 days, schisandra purifies the blood and sharpens the mind.

and wheezing. It may also be used to treat diarrhoea and dysentery, failing eyesight and hearing, as well as skin problems such as urticaria and eczema.

DRIED FRUIT

KEY INFORMATION

SAFETY	★★★★☆
TRADITIONAL USE	★★★★⯪
RESEARCH	★★★☆☆

BEST TAKEN AS Dried fruit or powder ✓✓✓ Tincture ✓✓

DOSAGE C (*see pp.44–45*)

OFTEN USED WITH Astragalus (*Astragalus chinensis*)

CAUTIONS Do not take during pregnancy. Can cause mild digestive irritation. See also pp.42–51.

BAICAL SKULLCAP

Scutellaria baicalensis

Used in both Chinese and Japanese herbal medicine, baical skullcap is a major remedy for allergic and inflammatory states. In traditional terms, it clears "hot and damp" conditions such as fever and dysentery; in the West, it is mainly used to treat asthma, hay fever, and allergies.

MEDICINAL USES

Part used Root

Key actions Anti-allergenic
● Antibacterial ● Anti-inflammatory

Allergies The herb can reduce the intensity of allergic reactions (usually in combination with other remedies) such as asthma, eczema, hay fever, and nettle rash. Best results are likely to be had when taken on professional advice.

Gastrointestinal problems A useful remedy for diarrhoea and gastrointestinal infection, baical skullcap is also helpful in upper digestive problems such as nausea and vomiting.

root has anti-inflammatory properties

FRESH ROOT

Other uses In China, the herb is used for respiratory infection, including colds, cough, and bronchitis, and figures in many prescriptions for high blood pressure. It appears to have anti-cancer properties and a *kampo* (traditional Japanese system of medicine) formula containing the herb is prescribed to support immune function in cancer. In the West, baical skullcap is prescribed for high blood pressure and chronic inflammatory diseases such as rheumatoid arthritis.

TINCTURE

A decoction made from the root can relieve chesty colds and wheeziness.

KEY INFORMATION

SAFETY	★ ★ ★ ★ ☆
TRADITIONAL USE	★ ★ ★ ★ ☆
RESEARCH	★ ★ ★ ☆ ☆

BEST TAKEN AS Tincture ✓✓✓
Capsule ✓✓ Decoction ✓
DOSAGE B *(see pp.44–45)*
CAUTIONS Very rarely, may cause side effects. See also pp.42–51.

ELDER

Sambucus nigra

Although poorly researched, elder is a safe and effective domestic remedy for ear, nose, and throat problems, its traditional use in Europe goes back to the earliest times. Both berries and flowers are collected to make wine, but a tincture or hot infusion of either will prove more therapeutically effective.

MEDICINAL USES

Parts used Flower ● Fruit

Key actions Anticatarrhal ● Anti-inflammatory ● Antioxidant ● Anti-viral ● Diaphoretic (stimulates sweating)

Common cold, flu, and fever

Elderflowers make an excellent cooling infusion for cold symptoms, flu-like colds, and mild feverish states, easing symptoms as well as countering infection. They combine well with yarrow (*Achillea millefolium*). Drink the infusion hot, sweetened with honey. Ripe elderberries contain high levels of vitamin C, and are strongly antioxidant. Take elderberry syrup or extract to counter infection and speed recovery. A recent clinical trial found that elderberry extract shortened recovery time in people suffering from influenza. In either form, the berries

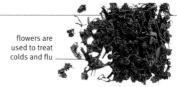

KEY INFORMATION

SAFETY	★ ★ ★ ★ ☆
TRADITIONAL USE	★ ★ ★ ★ ☆
RESEARCH	★ ★ ★ ☆ ☆

BEST TAKEN AS Flower: Infusion, tincture ✓✓✓ Fruit: Tincture, extract ✓✓✓
DOSAGE A (*see pp.44–45*)
OFTEN USED WITH Yarrow (*Achillea millefolium*)
CAUTIONS None known. Do not consume unripe berries. See also pp.42–51.

flowers are used to treat colds and flu

DRIED FLOWERS

can be taken to improve resistance to infection and reduce a tendency to recurring colds, sore throats, and coughs. Elderberry extract is commonly available as an over-the-counter remedy, and can safely be given to children.

Congestive problems of ear, nose, and throat

Elderflowers can dry and tone mucous membranes lining the nose and throat, reducing sneezing, itchiness, and a "runny nose" in conditions such as allergic rhinitis and hay fever. Combining elderflowers with nettle (*Urtica dioica*) is more effective in such cases. Elderflowers are often used in chronic catarrhal problems affecting the sinuses and the middle ear. In combination with other remedies, the

Found in temperate regions all over the world and often cultivated, the elder tree is native to Europe and thrives in woods and hedges, and on waste ground.

NATURE'S CURE-ALL

"If the medicinal properties of its leaves, bark and berries were fully known, I cannot tell what our countryman could ail for which he might not fetch a remedy from every hedge, either for sickness, or wounds. The buds boiled in water gruel have effected wonders in a fever; the spring buds are excellently wholesome in pattages; and small ale in which Elder flowers have been infused is esteemed by many...."
John Evelyn, 1664.

FLOWER CLUSTER

flowers are also useful in treating chest infections such as bronchitis and pleurisy.

Other uses Elderflower tea, taken hot rather than cold, is thought to act as a diuretic, stimulating urine flow. It is traditionally taken to relieve rheumatic aches and pains. A cold elderflower infusion can be used as a wash for sore and inflamed eyes, including conjunctivitis. Mildly astringent, it is believed to relieve skin conditions such as acne, eczema, and psoriasis.

The elder tree is traditionally known as "nature's medicine chest". The berries make excellent wines and winter cordials that relieve colds.

SKULLCAP

Scutellaria lateriflora

A key nerve tonic, skullcap is thought to have a "deeper" action on the nervous system than almost any other herb. Although poorly researched, it is used as a standard remedy for anxiety and nervous exhaustion, as well as related symptoms such as disturbed sleep, lowered mood, and headache.

MEDICINAL USES

Parts used Aerial parts

Key actions Antispasmodic ● Mild bitter ● Nerve tonic ● Sedative

Nervous tension and anxiety Skullcap is used primarily as a nerve tonic and a restorative. Taken on its own or in combination with other herbs, it soothes a tense and tired nervous system, and is helpful for headaches and migraine, an inability to relax, and poor sleep. Considered a "food" for the nervous system, skullcap often proves helpful in conditions where nervous debility is a factor, whether as the result of long-term stress, insomnia, or chronic pain. A range of other nerve-related disorders such as shock, dizziness, tinnitus, and chronic fatigue may also benefit. The herb's antispasmodic action makes it useful in relieving the taut and tensed muscles that so often accompany anxiety and worry. Skullcap is included in many over-the-counter herbal formulations for anxiety and sleep problems.

Premenstrual syndrome (PMS) Usually combined with chaste berry (*Vitex agnus-castus*) and taken in small doses throughout the menstrual cycle, skullcap can take the edge off symptoms of PMS such as oversensitivity, nervous irritability, and breast pain and tenderness. It may also relieve menstrual cramps.

Easily recognizable by its distinctive seed capsules and pink to blue flowers, skullcap is a native of the US and Canada.

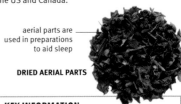

aerial parts are used in preparations to aid sleep

DRIED AERIAL PARTS

KEY INFORMATION

SAFETY	★★★★☆
TRADITIONAL USE	★★★★✮
RESEARCH	★✮☆☆☆

BEST TAKEN AS Tincture ✓✓✓ Capsule ✓✓ Tablet ✓
DOSAGE C (*see pp.44–45*)
OFTEN USED WITH Valerian (*Valeriana officinalis*)
CAUTIONS None known at normal dosage. See also pp.42–51.

CAPSULES

SAW PALMETTO

Serenoa repens

In many European countries, saw palmetto is used in the standard medical treatment for enlarged prostate gland. Strong research evidence supports its use in treating benign prostatic hypertrophy (BPH) and the symptoms that go with it.

MEDICINAL USES

Part used Fruit

Key actions Anti-inflammatory ● Prostate remedy ● Tonic for males

Prostate gland and urinary tract problems Best taken as standardized extract, saw palmetto helps to relieve mild to moderate BPH symptoms such as frequency and difficulty in passing urine, and improve prostate health. A useful anti-inflammatory within the urinary tract, it is beneficial in the treatment of chronic cystitis and urethritis.

diuretic and tonic in effect   **DRIED BERRIES**

Sexual tonic Saw palmetto has longstanding use as a male sexual tonic. However, in the 19th century, it was most prescribed as a female rather than a male sexual tonic, and it may well be that the herb improves libido in both men and women.

Polycystic ovary syndrome The extract is thought to have anti-androgenic activity and is used to treat polycystic ovary syndrome, a gynaecological condition in which androgen levels are raised.

A small palm tree, saw palmetto is native to Florida and the Gulf of Mexico, and grows in sand dunes along the coast.

KEY INFORMATION

SAFETY   ★ ★ ★ ★ ☆
TRADITIONAL USE   ★ ★ ★ ★ ☆
RESEARCH   ★ ★ ★ ★ ☆
BEST TAKEN AS Standardized extract ✓✓✓
DOSAGE C *(see pp.44–45)*
OFTEN USED WITH Nettle root (*Urtica dioica*)
CAUTIONS Occasionally may cause gastrointestinal upset, headache, or dizziness. See also pp.42-51.

MILK THISTLE

Silybum marianum syn. Carduus marianus

A powerful friend of the liver, milk thistle is thought to protect it against poisoning, toxicity, and inflammatory damage. Extracts have been shown to stimulate liver repair and regeneration, inhibit inflammatory processes resulting from toxins or infection, and promote effective liver detoxification.

MEDICINAL USES

Part used Seeds

Key actions Antioxidant ● Protects liver ● Stimulates breast milk

Liver and digestive disorders Milk thistle can prove very effective whenever the liver is under stress, usually reflected in raised liver enzyme levels. In liver disease and when taken to protect liver function during chemotherapy, it should be taken on professional advice. Long-term use appears to be safe. Liver

SEEDS help to maintain liver function

CAPSULE

conditions that can be treated with milk thistle are alcoholic hepatitis, liver cirrhosis, liver poisoning, and viral hepatitis. The seeds contain silymarin, a substance that protects the liver against poisoning, as for example, in the accidental ingestion of death cap mushrooms.

Other uses As its name implies, the seeds were taken by nursing mothers to improve the supply of breast milk, a use that remains applicable even today.

Native to the Mediterranean, milk thistle grows in the wild throughout Europe. It thrives mainly on waste ground, but is also cultivated as an ornamental plant.

KEY INFORMATION

SAFETY   ★ ★ ★ ★ ☆
TRADITIONAL USE   ★ ★ ★ ★ ★
RESEARCH   ★ ★ ★ ★ ☆
BEST TAKEN AS Standardized extract (standardized to 140mg silymarin) ✓ ✓ ✓
DOSAGE M, B (*see pp.44–45*)
CAUTIONS Occasionally may cause gastrointestinal upset and allergic reactions. See also pp.42-51.

SARSAPARILLA

Smilax spp.

Long used in the Americas and Europe, sarsaparilla is a valuable remedy for chronic infections, chronic inflammatory disease, and menopausal problems.

MEDICINAL USES

Parts used Root

Key actions Anti-inflammatory ● Anti-rheumatic ● Detoxicant ● Diuretic ● Tonic

Skin disorders Sarsaparilla is used to treat psoriasis and eczema, particularly where itchiness is a major factor. It is best taken in combination, for example with yellow dock (*Rumex crispus*).

Menopausal problems Sarsaparilla can help with menopausal problems linked with skin or arthritic symptoms.

FRESH LEAVES

KEY INFORMATION	
SAFETY	★ ★ ★ ★ ☆
TRADITIONAL USE	★ ★ ★ ★ ☆
RESEARCH	★ ★ ☆ ☆ ☆
BEST TAKEN AS	Decoction ✓✓✓ Tincture ✓✓
DOSAGE	C *(see pp.44–45)*
CAUTIONS	High doses may cause side effects. Seek professional advice if taking prescribed medication. See also pp.42–51.

BETONY

Stachys officinalis

Although poorly researched, betony has a wealth of traditional uses, with one classical writer recommending it for as many as 47 illnesses.

MEDICINAL USES

Parts used Aerial parts

Key actions Astringent ● Nerve tonic ● Mild bitter ● Mild sedative

KEY INFORMATION	
SAFETY	★ ★ ★ ★ ☆
TRADITIONAL USE	★ ★ ★ ☆ ☆
RESEARCH	★ ☆ ☆ ☆ ☆
BEST TAKEN AS	Infusion ✓✓✓ Tincture ✓✓ Tablet ✓
DOSAGE	C *(see pp.44–45)*
CAUTIONS	Do not take during pregnancy. See also pp.42–51.

Anxiety, nervous exhaustion and headache Betony tincture or infusion has particular application in chronic nervous states involving mental overactivity. It will help to relieve anxiety and irritability, as well as accompanying symptoms such as poor concentration and headache. It may also be taken for dizziness and nerve pain.

FLOWERS

A European herb, milk thistle (*Silybum marianum*) has white markings on its leaves caused, folklore has it, by the Virgin Mary's milk. It is today the foremost herbal remedy for liver problems.

COMFREY

Symphytum officinale

Comfrey root is highly effective in stimulating tissue repair. When applied regularly to damaged tissue such as sprains, bruises, sports injuries, and operation scars, it promotes regrowth and shortens recovery or repair time. Comfrey ointment or cream merits a place in every home first aid kit.

MEDICINAL USES

Parts used Leaf ● Root

Key actions Anti-inflammatory ● Astringent ● Demulcent ● Wound healer

Bruises, sprains, and tissue repair

Applied as soon as possible to the site of bruises, sprains, or minor fractures, comfrey cream or ointment will often minimize swelling and promote quick and effective repair. Continue applying, as ointment, cream, or poultice of leaves and root, until tissue is healed. Comfrey also helps with varicose veins, slow-healing wounds and ulcers. Where the wound is still open, apply comfrey carefully around the margins of the wound, not directly on it. Apply comfrey with caution during pregnancy.

Comfrey, which is also known as "knitbone", stimulates bone repair.

KEY INFORMATION

SAFETY	★★★★☆
TRADITIONAL USE	★★★★★
RESEARCH	★★★☆☆

BEST TAKEN AS Ointment ✓✓✓ Leaves, cream, lotion ✓✓
DOSAGE T, Leaf: C (*see pp.44–45*)
OFTEN USED WITH Yarrow (*Achillea millefolium*)
CAUTIONS Do not take comfrey root internally. Do not take comfrey leaf internally during pregnancy, or for more than 6 weeks at a time. Do not apply comfrey to open wounds. Restricted herb in some countries; internal use not legal in Australia. See also pp.41–52.

root helps to heal bruises

FRESH ROOT

CHICKWEED

Stellaria media

Best known as a remedy for itchy skin, chickweed can also bring relief to problems such as eczema, nettle rash, and irritated varicose veins.

MEDICINAL USES

Part used Whole plant

Key actions Astringent ● Cooling (topically) ● Demulcent ● Relieves itchiness

Itchy skin The cream and freshly squeezed juice are markedly cooling on the skin and can be applied as often as wanted to soothe sore and itchy areas. The infusion can be added to a bath or cooled and used as a wash on varicose veins.

CREEPING PLANT

> **KEY INFORMATION**
>
> SAFETY ★★★☆☆
> TRADITIONAL USE ★★★★☆
> RESEARCH ★☆☆☆☆
> BEST TAKEN AS Cream ✓✓✓ Freshly squeezed juice ✓✓
> DOSAGE T *(see pp.44–45)*
> CAUTIONS Can cause allergic skin reactions; try a small quantity first. See also pp.42–51.

STEVIA

Stevia rebaudiana

Stevia is a non-sugar natural sweetener that lowers blood sugar levels and makes a good replacement for both sugar and artificial sweeteners.

MEDICINAL USES

Part used Leaf

Key actions Antimicrobial ● Hypoglycaemic (lowers blood sugar levels) ● Lowers blood pressure

Sugar replacement The herb's sweet taste and hypoglycaemic action make it a valuable remedy in early onset diabetes. It can also help to prevent tooth decay, aid weight loss, and improve immune resistance in yeast infections. Take on its own as a tea or use in place of sugar – $1/4$ teaspoon of ground leaf is roughly equivalent to 1 teaspoon of sugar.

leaves are sweet-tasting

FRESH LEAVES

> **KEY INFORMATION**
>
> SAFETY ★★★★☆
> TRADITIONAL USE ★★★★☆
> RESEARCH ★★★★☆
> BEST TAKEN AS Infusion ✓✓
> DOSAGE Food, C *(see pp.44–45)*
> CAUTIONS Avoid excessive doses in pregnancy. See also pp.42–51.

FEVERFEW

Tanacetum parthenium

A cooling, bitter remedy that was once used to treat fever, feverfew is now a standard remedy for tension headache and migraine. Research, which has by and large confirmed the herb's effectiveness, began after a Welsh doctor's wife found feverfew cured her 50-year-long history of migraine.

MEDICINAL USES

Parts used Leaf

Key actions Anti-inflammatory ● Bitter

Headache and migraine Feverfew proves most effective when it is taken as soon as signs of an impending migraine attack are recognized. The herb is generally less effective once the migraine is underway. The leaf can be used symptomatically, for example in tension headache. But for best results, the herb should be taken regularly for several months. Strong-acting and potentially toxic, feverfew should be taken at the recommended dosage – either one small fresh leaf with food or as a capsule or tablet produced to pharmacopoeial standards.

Arthritic pain The herb's anti-inflammatory action is useful in treating arthritic aches and pain. It is most likely to relieve arthritic pain when combined with herbs such as willow bark (*Salix alba*) or devil's claw (*Harpagophytum procumbens*).

dried leaves are taken to relieve migraine

DRIED LEAVES

KEY INFORMATION

SAFETY	★ ★ ★ ⯪ ☆
TRADITIONAL USE	★ ★ ★ ★ ☆
RESEARCH	★ ★ ★ ⯪ ☆

BEST TAKEN AS Tablet ✓✓✓ Capsule ✓✓ Fresh leaf ✓

DOSAGE Tablet, capsule: M (*see pp.44–45*) one small fresh leaf a day.

CAUTIONS Avoid during pregnancy or while breast-feeding. If taking blood-thinning medication, take only on the advice of a herbal or medical practitioner. Can cause allergic reactions, mouth ulcers, and gastrointestinal upset. See also pp.42–51.

Other uses Feverfew has many traditional uses and has often been used to treat hot and feverish states. A strong bitter, it stimulates appetite and digestive activity and has been employed to treat worms. The leaf acts to stimulate menstrual blood flow and can prove helpful in relieving menstrual pains.

DECOCTION

Feverfew has white and yellow daisy-type flowers, and care must be taken not to confuse chamomile (*Chamomilla recutita*) with this potentially toxic herb.

PAU D'ARCO, LAPACHO

Tabebuia spp.

A traditional South American remedy, pau d'arco is thought to be a specific for problems such as yeast infection and candidiasis – applied to the skin or taken internally.

MEDICINAL USES

Part used Bark

Key actions Antibacterial ● Antifungal ● Astringent ● Immune-stimulant ● Reputed anti-tumour activity

Fungal and bacterial infection Best taken in combination with other herbs such as echinacea (*Echinacea* spp.)

PAU D'ARCO TREE

or golden seal (*Hydrastis canadensis*) to treat digestive infections, tonsillitis, thrush, and candidiasis, it is used in Brazil as a preventative and adjuvant in cancer treatment (unsupported by research).

KEY INFORMATION

SAFETY   ★ ★ ★ ⯪ ☆
TRADITIONAL USE   ★ ★ ★ ★ ★
RESEARCH   ★ ★ ★ ☆ ☆
BEST TAKEN AS Decoction ✓✓✓
Tincture ✓✓ Capsule ✓
DOSAGE C *(see pp.44–45)*
CAUTIONS Do not take during pregnancy. If on prescribed anticoagulant medication, take only on the advice of a herbal or medical practitioner. See also pp.42–51.

TAMARIND

Tamarindus indica

Best known for its tart, slightly spicy flavour in chutneys and sauces, tamarind is mainly used as a gentle laxative to treat constipation in children.

MEDICINAL USES

Part used Fruit

Key actions Laxative ● Nutritive

Constipation Fresh or dried fruit can be made into a pleasant drink and taken to help open the bowels and relieve constipation.

Other uses Recent research has found that the fruit increases the availability of ibuprofen in the body, suggesting that it could be used in arthritis to reduce the dosage of aspirin-type medicines.

TAMARIND TREE

KEY INFORMATION

SAFETY   ★ ★ ★ ★ ⯪
TRADITIONAL USE   ★ ★ ⯪ ☆ ☆
RESEARCH   ★ ★ ☆ ☆ ☆
BEST TAKEN AS Fruit ✓✓✓
DOSAGE Food, M *(see pp.44–45)*
CAUTIONS May interact with aspirin-type medicines. See also pp.42–51.

DANDELION

Taraxacum officinale

One of nature's most versatile remedies, dandelion is both a nutritious salad vegetable and a detoxifying remedy for the liver and kidneys. Loved by herbalists for its gentle cleansing effect, dandelion root finds use in toxic states of all kinds, including chronic skin disorders and recurrent infection.

MEDICINAL USES

Parts used Root ● Leaf

Key actions Bitter tonic ● Diuretic ● Liver cleanser ● Mild laxative

Poor appetite and digestion, poor liver function Bitter but not excessively so, dandelion root has a beneficial action on the stomach, liver, and pancreas, increasing digestive secretions, including bile, and tending to stabilize blood sugar levels. The root promotes liver detoxification.

Fluid retention, high blood pressure The standard herbal diuretic, dandelion leaf acts mainly on the kidneys and encourages fluid clearance and weight loss. It is commonly taken to help lower blood pressure, its high potassium content making it particularly useful.

Skin problems A gentle "blood cleanser", dandelion root will prove helpful in a number of chronic skin complaints such as acne, boils, and eczema, especially when it is combined with herbs such as burdock *(Arctium lappa)* and echinacea *(Echinacea* spp.*)*. The leaves are cleansing and nutritious, and make a good addition to salads.

A good diuretic, dandelion is also known as piss-a-bed.

DRIED ROOT | root gently stimulates appetite

KEY INFORMATION

SAFETY ★ ★ ★ ★ ★
TRADITIONAL USE ★ ★ ★ ★ ★
RESEARCH ★ ★ ★ ☆ ☆
BEST TAKEN AS Root: Tincture ✓✓✓
Leaf: Infusion ✓✓✓ Tincture ✓✓
DOSAGE A (*see pp.44–45*)
OFTEN USED WITH Burdock (*Arctium lappa*)
CAUTIONS Occasionally can cause allergic reaction. See also pp.42–51.

CACAO, CHOCOLATE

Theobroma cacao

For the Maya, cacao or chocolate was "the food of the gods". More a food than a medicine, it offers distinct health benefits. Traditionally used as a heart and kidney tonic, dark chocolate is today recommended as an antioxidant to support cardiovascular health.

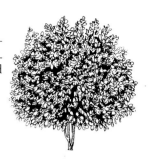

MEDICINAL USES

Part used Seeds

Key actions Antioxidant ● Diuretic ● Mild bitter ● Nutritive ● Stimulant

Mood enhancement Due in part to an influence on serotonin and endorphin levels, cacao induces subtle effects on the mind and emotions, increasing alertness while calming and relaxing the body. Moderate amounts of plain chocolate will enhance mood and support a positive mental state. Overall, cacao induces a sense of well-being, and as part of a broad approach, can help to lift lowered mood, especially when linked to nervous exhaustion. Where it is difficult to maintain a moderate intake, for example in pre-menstrual sugar-craving, other remedies such as damiana (*Turnera diffusa*) or St John's wort (*Hypericum perforatum*) may be more appropriate.

Applied locally, the "butter" extracted from cacao beans helps to nourish and protect the skin and mucous membranes.

Other uses A good addition to the diet to maintain cardiovascular health, cacao or dark chocolate at the end of a meal stimulates digestive activity. Its polyphenols exert an antioxidant activity within the heart and stomach. Research has shown that cacao helps counter bacteria that cause septicaemia and boils. Cacao butter is also widely used in making cosmetic preparations, lip salves, and pessaries.

beans contain a nutritious oil

BEANS IN A POD

KEY INFORMATION

SAFETY ★ ★ ★ ★ ★
TRADITIONAL USE ★ ★ ★ ★ ☆
RESEARCH ★ ★ ★ ★ ☆
BEST TAKEN AS Infusion ✓✓✓
Plain chocolate ✓✓
DOSAGE Food
CAUTIONS Can cause migraine headache and gastrointestinal upset. See also pp.42–51.

roasted beans are used to make chocolate

ROASTED BEANS

Growing wild in most parts of the world, dandelion (*Taraxacum officinale*) leaf makes a nutritious addition to salads. Rich in vitamins A and C and iron, the leaf supports liver and kidney function.

THUJA, TREE OF LIFE

Thuja occidentalis

Native Americans employed thuja for conditions such as headache, fever, and rheumatism, and burnt it as a cleansing "smudge". Thuja's key application is for warts, though it helps in other infectious conditions as well – notably sinusitis, tooth abscesses, bronchitis, cystitis, and fungal infections.

MEDICINAL USES

Part used Leaf

Key actions Antifungal ● Antimicrobial ● Anti-viral ● Blood cleanser

Warts and topical application No remedy is guaranteed to remove warts, but thuja is more likely to succeed than many others. Apply a drop or two of neat tincture to the wart twice a day. Continue for up to 10 days.

Infections Not normally taken on its own, thuja combines well with other antimicrobial and immune-enhancing remedies, notably echinacea (*Echinacea* spp.) and thyme (*Thymus vulgaris*). Its marked antiseptic activity is most apparent in viral and bacterial infections affecting mucous membranes, especially membranes within the ear, nose, throat, and urinary tract. Strong-acting and potentially toxic when taken internally, thuja is best used on professional advice.

LEAVES

Other uses Thuja is prescribed by practitioners for a wide variety of conditions, including psoriasis, fibroids, and bed-wetting. It has been prescribed for uterine polyps and uterine cancer, which can be caused by the wart virus that produces polyps. In rheumatic problems, a lotion can be applied topically as a counter-irritant to relieve muscular aches and pains.

Thuja contains a strong volatile oil with potent antifungal and anti-viral activity.

KEY INFORMATION

SAFETY	★ ★ ★ ☆ ☆
TRADITIONAL USE	★ ★ ★ ★ ☆
RESEARCH	★ ★ ★ ☆ ☆
BEST TAKEN AS	Tincture ✓✓✓

DOSAGE D (*see pp.44–45*)

OFTEN USED WITH Echinacea (*Echinacea* spp.)

CAUTIONS Do not take thuja during pregnancy and while breast-feeding. See also pp.42–51.

THYME, COMMON THYME

Thymus vulgaris

A classic kitchen herb, thyme makes a refreshing tea that counters infection and tones the respiratory system. Useful in almost any problem affecting the ear, nose, throat, and chest, thyme disinfects the air passages, soothes coughing, and stimulates clearance of phlegm.

MEDICINAL USES

Parts used Aerial parts

Key actions Antibacterial ● Antifungal ● Antioxidant ● Expectorant ● Relaxant ● Tonic

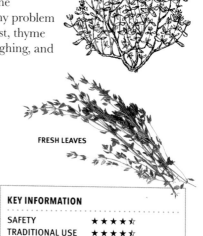

FRESH LEAVES

Ear, nose, and throat (ENT) problems Thyme tea, with or without a spoonful of honey, is an excellent home remedy for ENT problems, including colds, catarrh, sinus congestion, sore throat, and tonsillitis. The tea can be first used as a gargle and then swallowed.

Cough and bronchial infection Thyme brings relief to all manner of coughs and chesty problems, and can provide valuable support in asthma and whooping cough. It is often combined with liquorice (*Glycyrrhiza glabra*) and echinacea (*Echinacea* spp.).

Other uses The tea may be taken as a general tonic, as well as to relieve indigestion and wind and to treat threadworms. In arthritic and rheumatic conditions, it makes an invigorating addition to a bath. The essential oil can be applied neat to fungally infected nails; use 1 drop per nail twice a day (do not use neat elsewhere and do not take internally).

KEY INFORMATION

SAFETY ★ ★ ★ ★ ☆

TRADITIONAL USE ★ ★ ★ ★ ☆

RESEARCH ★ ★ ★ ☆ ☆

BEST TAKEN AS Infusion ✓✓✓
Tincture ✓✓

DOSAGE C *(see pp.44–45)*

OFTEN USED WITH Liquorice (*Glycyrrhiza glabra*)

CAUTIONS Rarely, may cause gastrointestinal upset or allergic reaction. See also pp.42–51.

DRIED AERIAL PARTS

Traditionally, thyme is seen as a "longevity" herb, a view supported by evidence that it prevents the breakdown of essential fatty acids within the brain.

LIMEFLOWER, LINDEN BLOSSOM

Tilia spp.

Commonly planted in gardens and parks, the lime or linden tree bears flowers that perfume the air on summer evenings. The delicate scent acts in much the same way as the infusion made from the flowers, soothing troubled states of mind and relieving tension headache, migraine, and sinus congestion.

MEDICINAL USES

Part used Flower

Key actions Antispasmodic
● Diaphoretic (stimulates sweating)
● Mild sedative ● Nerve tonic

Colds, catarrh, and fever Limeflower is considered to be a first-rate remedy for head colds and mild fevers, as well as nasal or sinus catarrh; drink small amounts of limeflower tea frequently throughout the day. It is an excellent remedy for children, if need be blended with some apple juice in order to improve the taste. The tea also makes a good steam inhalation to ease sinus headache and congestion. Limeflower combines well with elderflower (*Sambucus nigra*).

Native to Europe, the lime tree is found in the wild. These trees grow up to 30m (100ft), with heart-shaped leaves and clusters of yellow flowers.

Anxiety and tension Under-appreciated, perhaps because it is a gentle-acting remedy, limeflower has constituents that exert a mild tranquillizing effect, similar in some respects to benzodiazepine tranquillizers (for example, Xanax, Valium). Usually best taken as an infusion or tincture, limeflower helps to cool and relieve anxiety states, particularly when symptoms include head and neck tension, palpitations, and feeling "hot and bothered". Its mild action makes it valuable in helping to soothe agitation and restlessness in children, and it can also be taken to calm emotional shock, combining well here with oat straw (*Avena sativa*).

DRIED FLOWERS

Poor sleep A good night-time drink, limeflower tea is safe for children and adults alike, encouraging relaxation and a drift into sleep rather than being overly sedative. Combine with passion flower (*Passiflora incarnata*) to increase the sedative strength of the tea. For nervous tension and to aid sleep, 50–300g

KEY INFORMATION

SAFETY	★ ★ ★ ★ ✬
TRADITIONAL USE	★ ★ ★ ★ ☆
RESEARCH	★ ★ ★ ☆ ☆
BEST TAKEN AS	Infusion ✓ ✓ ✓
DOSAGE	B *(see pp.44–45)*
OFTEN USED WITH	Hawthorn

(*Crataegus* spp.)
CAUTIONS Can occasionally cause allergic reactions. Pollen can provoke hay fever. See also pp.42–51.

The father of modern botany and ecology, Carl Linné or Linnaeus (1707–78) reputedly owed his name to the lime tree that grew by his family home.

(2–10oz) of flowers can be infused for 20 minutes in the evening and added to a warm bath. Relax and soak in the bath, then retire to bed.

Palpitations and high blood pressure

A specific for nervous palpitations, the flowers are thought to slow and stabilize the heart rate and rhythm and are often prescribed by herbal practitioners for an irregular or racing heartbeat. Limeflower can also be valuable as part of a broad approach to treating high blood pressure, especially where this is associated with arteriosclerosis (hardening of the arteries) and nervous tension. Taken long term, limeflower's high bio-flavonoid content helps to improve the health of the arteries, and the flowers combine particularly well with hawthorn (*Crataegus* spp.) to support healthy heart function and circulation.

Other uses Limeflower stimulates blood flow to the capillaries and surface of the body, thereby stimulating sweating and

> ### LIMEFLOWER HONEY
>
> "Bees are extremely fond of flowers of the lime, which abound with honey, and they also sometimes collect the sweet 'honey dew', the deposit of aphids, that covers the leaves during summer. In Lithuania, holes are made in large trees, which the bees soon convert into hives and the combs are removed when full. Such honey is thought a valuable remedy in Poland for lung disorders, and fetches a very high price."
> C. Pierpoint Johnson, 1875
>
> **HONEY BEE**

helping to cool the body in hot and feverish states. When combined with circulatory stimulants such as ginkgo (*Ginkgo biloba*) or angelica (*Angelica archangelica*), this circulatory effect can help to improve peripheral circulation and a tendency to cold extremities. Conditions such as chilblains and restless legs can also benefit.

PUNCTURE VINE, TRIBULUS

Tribulus terrestris

This plant has figured for at least 2,000 years in both Western and Eastern traditions as a remedy for kidney and bladder problems, particularly kidney stones. It also possesses a long-standing reputation as a tonic and aphrodisiac.

MEDICINAL USES

Parts used Leaf ● Fruit ● Root

Key actions Anti-inflammatory ● Oestrogenic ● Reputed aphrodisiac ● Stimulates menstruation

Kidney and urinary tract problems

A useful remedy for urinary problems such as cystitis and urethritis, puncture vine can be particularly helpful for the chronic urethral irritation that quite often occurs as oestrogen levels fall at the time of the menopause. In combination with other remedies, it has been commonly used to help clear kidney and bladder stones, though for these conditions it needs to be taken on professional advice.

Puncture vine thrives on waste land. A thorny, creeping plant, it can be found growing all the way from south-eastern Europe to China.

Sexual tonic A tonic for both men and women, puncture vine is likely to help most where sexual vitality and libido are at a low ebb. In women, it can help to improve libido, especially during the menopause, and in men, there are indications that it can help with problems such as erectile dysfunction and lowered testosterone levels.

Body building Currently much in vogue as a body-building supplement, puncture vine contains steroidal saponins, which

are thought to increase muscle bulk in much the same way as testosterone. Despite the hype surrounding the herb, the root and leaf clearly do have hormonal activity within the body, though they are probably only effective (anabolic) when testosterone levels are low.

fruit contains steroidal saponins

CAPSULES   **FRUIT**

KEY INFORMATION

SAFETY ★ ★ ★ ⯪ ☆
TRADITIONAL USE ★ ★ ★ ★ ☆
RESEARCH ★ ★ ☆ ☆ ☆
BEST TAKEN AS Tablet ✓✓✓ Capsule ✓✓
DOSAGE M (*see pp.44–45*)
CAUTIONS Keep to recommended dosage. Avoid in pregnancy. See also pp.42–51.

FENUGREEK

Trigonella foenum-graecum

Strongly mucilaginous, fenugreek soothes and heals sore, inflamed, or ulcerated tissue in the gastrointestinal tract. Clinical trials have shown that it lowers levels of "bad" fats – cholesterol, LDL, VLDL, and triglycerides – within the blood. Blood sugar control and insulin response in diabetics are also improved.

MEDICINAL USES

Parts used Seed • Sprouts

Key actions Anti-inflammatory • Demulcent • Expectorant • Laxative • Lowers cholesterol and blood sugar levels

Gastrointestinal problems Inflammatory problems within the digestive tract such as mouth ulcers, gastritis, and irritable bowel will benefit from the seeds' moistening and protective effect. For best results, first soak the seeds in water.

Cholesterol and blood sugar Clinical evidence supports fenugreek's use in raised cholesterol levels; however, large doses were used in the clinical trials (5–100g a day) to achieve this result. The seeds can also be used on a daily basis to help reduce blood sugar levels in diabetes. Components in fenugreek stimulate insulin as well as delay absorption of glucose.

Fenugreek seeds are strongly oestrogenic and can be valuable in menopausal symptoms.

seeds have a demulcent action

SEEDS

KEY INFORMATION

SAFETY	★ ★ ★ ★ ☆
TRADITIONAL USE	★ ★ ★ ★ ★
RESEARCH	★ ★ ★ ☆ ☆

BEST TAKEN AS Ground or sprouting seeds ✓✓✓ Decoction ✓✓ Capsule ✓
DOSAGE A (*see pp.44–45*)
CAUTIONS In pregnancy, or if taking prescribed anticoagulant or diabetic medication, take only on the advice of a herbal or medical practitioner. Occasionally can cause gastrointestinal upset. See also pp.42–51.

RED CLOVER

Trifolium pratense

Traditionally seen as a "hot" and "dry" herb, red clover's dense red flowers were thought in medieval times to signify its value as a blood cleanser. Taken to clear chronic toxicity, red clover is an important remedy for skin problems such as acne, boils, eczema, and psoriasis.

MEDICINAL USES

Part used Flower

Key actions Antispasmodic
● Blood cleanser ● Expectorant
● Phytoestrogenic ● Reputed anti-cancer activity

Chronic toxicity A mild laxative, red clover is most likely to aid detoxification in cases where skin and glandular problems are linked with chronic constipation. Safe for children with mild skin conditions or swollen glands linked to a sore throat, red clover works best with small doses initially that are slowly increased. It combines well with other skin remedies, including echinacea (*Echinacea* spp.). The herb's use as an anti-cancer remedy is unsubstantiated, though its ability to stimulate cleansing of the lymph system suggests that it may have a role as an adjuvant along with herbs such as yellow dock (*Rumex crispus*). Red clover also works well with herbs such as marigold (*Calendula officinalis*) to ease swollen and painful breasts, but concentrated extracts

INFUSION | infusion supports skin health

should not be taken during pregnancy and while breast-feeding.

Chest problems Traditionally given to children as a remedy for chesty coughs and wheeziness, red clover can prove useful in chest problems, especially when combined with thyme (*Thymus vulgaris*). Taken as a hot tea sweetened with honey, this combination can help to soothe chronic and irritable coughs and may be helpful alongside prescribed medication for problems such as bronchitis and bronchial asthma. Red clover has also been used to treat night sweats associated with chest infection.

KEY INFORMATION

SAFETY	★ ★ ★ ★ ⯪
TRADITIONAL USE	★ ★ ★ ★ ☆
RESEARCH	★ ★ ☆ ☆ ☆

BEST TAKEN AS Tincture ✓✓✓
Infusion ✓✓ Capsule, tablet ✓
DOSAGE A *(see pp.44–45)*
OFTEN USED WITH Yellow dock (*Rumex crispus*)
CAUTIONS Rarely, can cause headache or skin rash. Do not take concentrated extracts in pregnancy and while breast-feeding. See also pp.41–52.

flowers promote detoxification

DRIED FLOWERS

THE CLOVER LEAF

Many classical stories and myths relate to the clovers. Perhaps the most enduring legacy of these myths concerns the "club" of our playing cards, said to resemble a clover leaf. The clover leaf in turn resembles the three-lobed club wielded by Hercules, known in Latin as *clava trinodis*.

ACE OF CLUBS

Menopausal symptoms Concentrated isoflavone extracts that are strongly oestrogenic are available over the counter as an alternative form of hormone replacement therapy. They can prove helpful in relieving menopausal symptoms that are usually linked to lowered oestrogen levels, such as hot flushes, night sweats, headaches, and poor sleep. Due to the high levels of isoflavone, concentrated extracts also have anti-inflammatory activity and support the health of the heart and circulation. However, concentrated extracts such as these are very different from typical herbal preparations, and need to be seen as separate products with distinct areas of activity. In view of the high levels of phyoestrogens present, it is advisable to take concentrated isoflavone extracts up to a maximum of 3 months at a time. Where repeated use is wanted, it is best to seek professional advice.

FRESH HERB

Other uses A lotion made from the tea makes a useful skin wash for persistent sores, inflamed skin, and swollen insect bites. The flowers may also be decocted, strained and while warm applied as a poultice on swollen and tender glands.

In Spain, red clover was traditionally used to help in treating cataracts. The pale crescent marking on the herb's leaves was thought to resemble a cataract.

A common wayside and meadow plant, red clover (*Trifolium pratense*) has pink, oval-shaped flowerheads, which in the past were gathered and used for dyeing wool green.

DAMIANA

Turnera diffusa var. *aphrodisiaca*

Used for centuries by the Maya of central America, damiana continues to be a popular herb throughout central and south America, and is taken as a nerve tonic and aphrodisiac. It is thought to improve vitality and libido in both men and women, though it is unclear quite how effective it is in this respect.

MEDICINAL USES

Part used Leaf

Key actions Mild antidepressant
● Nerve tonic ● Aphrodisiac ● Diuretic

Nerve tonic and aphrodisiac
Taken regularly damiana will help to improve mood and mental stamina. It is often most useful where anxiety and depression occur together, a state which frequently leads to (and can be caused by) nervous exhaustion. In the Anglo-American herbal tradition, damiana is a nervine and a specific for depleted nervous states, often being combined with St John's wort (*Hypericum perforatum*). It can also prove useful in treating headaches, especially when linked to the menstrual cycle.

Damiana was used by the Maya as a tonic and as a remedy for giddiness and loss of balance.

Aphrodisiac Damiana clearly acts as a sexual tonic, though whether it is more effective than other sexual tonics, such as puncture vine (*Tribulus terrestris*), is debatable. Traditionally it has been seen as a male tonic, but it can equally help to improve sexual activity in women.

Other uses Damiana is a diuretic and has antiseptic constituents that make it potentially useful in urinary tract infection. It has been used to treat prostatitis. The herb is also mildly laxative.

KEY INFORMATION	
SAFETY	★ ★ ★ ★ ☆
TRADITIONAL USE	★ ★ ★ ★ ☆
RESEARCH	★ ★ ✫ ☆ ☆
BEST TAKEN AS	Infusion ✓✓✓
Tincture ✓✓ Capsule ✓	
DOSAGE B (*see* pp.44–45)	
CAUTIONS None known. See also pp.42–51.	

SLIPPERY ELM

Ulmus rubra

Few herbs are more valued in Anglo-American herbal medicine than slippery elm. Readily digestible, the powdered bark can be mixed with water to form a thick jelly-like solution that has limitless applications, whether topically on the skin or internally within the gastrointestinal tract.

MEDICINAL USES

Part used Root (powdered)

Key actions Antioxidant ● Demulcent ● Emollient ● Nutritive

Digestive and respiratory problems

Slippery elm soothes and protects in problems such as heartburn, irritable bowel, and bronchitis. Stir 1–2 teaspoonfuls of powder into water and leave to stand for 5 minutes before drinking. Add a pinch of cinnamon powder, if wanted. Repeat as desired. Be sure to buy slippery elm powder, and not wheat powder with added slippery elm. Take slippery elm at least 30 minutes apart from other medications, as it tends to reduce absorption.

Topical uses As a "drawing" poultice for splinters, boils, and ulcers, mix a small quantity of slippery elm with echinacea (*Echinacea* spp.) infusion or tincture to form a thick paste. Spread onto the affected area and bandage. Leave for 24 hours. Repeat as required.

KEY INFORMATION

SAFETY	★ ★ ★ ★ ★
TRADITIONAL USE	★ ★ ★ ★ ★
RESEARCH	★ ★ ★ ☆ ☆

BEST TAKEN AS Powdered root (mixed with water) ✓✓✓

DOSAGE A *(see pp.44–45)*

OFTEN USED WITH Echinacea (*Echinacea purpurea*)

CAUTIONS Use only inner bark. Mix powder with plenty of water. Rarely, may cause local irritation. See also pp.42–51.

powdered root soothes inflammation

POWDERED ROOT

Slippery elm coats the inner lining of the stomach and intestines, relieving acidity, irritability, and inflammation.

CAT'S CLAW, UNA DE GATO

Uncaria tomentosa

A climbing vine native to the Amazon rainforest, cat's claw is believed by local healers to have great medicinal virtue; they use it as a cure-all to treat everything from asthma and arthritis to diabetes and cancer. It is an endangered species in the wild; use organic products only.

MEDICINAL USES

Part used Stem bark

Key actions Anti-inflammatory
● Antioxidant ● Immune tonic

Chronic infection Cat's claw's tonic action on the immune system makes it a key remedy for chronic infection and degenerative diseases. Best taken combined with other immune-enhancing herbs, cat's claw can prove useful in treating chronic fatigue, fibromyalgia, glandular fever, and herpes infection. Clinical studies in Peru suggest the herb can be helpful in HIV infection. It is a first-rate convalescent herb.

DRIED BARK

Anti-inflammatory Cat's claw has potent anti-inflammatory activity and can be successfully used to treat problems such as gastric ulceration, as well as

Cat's claw helps to boost the immune system, and is also useful in gastrointestinal infection and inflammation.

inflammatory joint problems such as rheumatoid arthritis and osteoarthritis.

Anti-cancer remedy Reflecting its traditional South American use, cat's claw has anti-tumour properties that make it valuable as an adjuvant treatment in cancer. Take with other appropriate herbal medicines and on professional advice.

THE "CAT'S CLAW"

KEY INFORMATION	
SAFETY	★ ★ ★ ⯪ ☆
TRADITIONAL USE	★ ★ ★ ★ ★
RESEARCH	★ ★ ★ ☆ ☆
BEST TAKEN AS	Tincture ✓✓✓ Tablets (preferably organic standardized) ✓✓
DOSAGE	B *(see pp.44–45)*
OFTEN USED WITH	Echinacea *(Echinacea purpurea)*
CAUTIONS	Do not take during pregnancy and while breast-feeding. May have contraceptive activity. Take on the advice of a herbal or medical practitioner if on blood-thinning or immunosuppressant medication. Can cause nausea or headache. See also pp.42–51.

CRANBERRY

Vaccinium macrocarpon

A well-known household remedy, cranberry is commonly taken for urinary tract problems such as cystitis and urethritis. Sharp-flavoured and rich in vitamin C, it has strong disinfectant properties within the urinary and gastrointestinal tracts.

MEDICINAL USES

Parts used Fruit • Juice

Key actions Antioxidant • Antiseptic

Cystitis and urethritis Cranberry contains constituents that make it difficult for bacteria to cling to the wall of the urinary tubules and intestines so that harmful bacteria are more readily flushed out of the body. Best taken as unsweetened juice or concentrated extract, cranberry can be combined with other fruit juices, for example apple juice, to make it more palatable. Avoid juices with large amounts of added sugar, not least because sugar depresses the immune system. Clinical research suggests that large quantities can be drunk for acute infection – up to ¾ litre (26 fl oz) a day for a few days, along with plenty of water. For long-term use, take less than half this quantity.

juice flushes out bacteria

JUICE

The North American cranberry has been well researched and is now recognized as a safe and effective treatment for mild urinary tract infection.

Prostate problems Cranberry's tonic and antiseptic action within the urinary tract makes it useful in problems affecting the prostate gland. Taken regularly as a juice or extract, it can help to ease frequency and other symptoms associated with an enlarged prostate. Other prostate problems, such as chronic prostatitis, may benefit from medium to long-term use.

Other uses Cranberry can also be a valuable supplement to take in digestive infections and diarrhoea, and in allergic states affecting the gut or respiratory system.

BERRIES

berries are strongly antioxidant

KEY INFORMATION

SAFETY	★ ★ ★ ★ ✬
TRADITIONAL USE	★ ★ ★ ✬ ☆
RESEARCH	★ ★ ★ ★ ☆

BEST TAKEN AS Unsweetened juice ✓✓✓ Fruit ✓✓ Tablet ✓

DOSAGE A *(see pp.44–45)*

OFTEN USED WITH Buchu *(Barosma betulina)*

CAUTIONS Very high doses may increase effect of prescribed anticoagulant medication, and can cause gastrointestinal upset. See also pp.42–51.

NETTLE

Urtica dioica, U. urens

Infamous for its stinging leaves, nettle is a fine example of a weed that has great value as food and medicine. Rich in iron, calcium, and silica, nettle leaf makes a useful tonic food – as tea or soup – in anaemia and rheumatic problems. Nettle leaf also has marked anti-allergenic activity.

MEDICINAL USES

Parts used Aerial parts ● Root

Key actions Anti-allergenic ● Anti-inflammatory ● Antispasmodic (root) ● Blood cleanser ● Diuretic ● Tonic (leaf)

Arthritic and rheumatic problems
Nettle's primary use is for stiff and inflamed joints, with conditions such as gout benefiting especially. Taken long-term for arthritic and rheumatic symptoms, nettle leaf can relieve pain and inflammation and support tissue repair. A safe remedy, nettle leaf tea or soup can be taken in relatively large amounts to maximize its anti-inflammatory activity. In a German clinical trial, patients with osteoarthritis were able to significantly reduce their intake of aspirin-type anti-inflammatories on taking nettle leaf. High levels of histamine and serotonin in the stinging hairs are thought to be responsible for the herb's "sting", and may explain the ancient practice of flaying an arthritic joint with nettle leaves (*U. urens*) in order to treat pain and

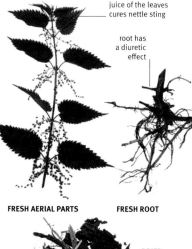

juice of the leaves cures nettle sting

root has a diuretic effect

FRESH AERIAL PARTS   **FRESH ROOT**

DRIED AERIAL PARTS

stiffness. This traditional use of nettle received unexpected confirmation in a recent double-blind clinical trial.

Skin disorders A good detox remedy, nettle leaf combines well with other herbs such as calendula (*Calendula officinalis*) and yellow dock (*Rumex crispus*) to treat all manner of chronic skin problems such as eczema, psoriasis, and urticaria. Apply nettle infusion as a lotion to relieve inflamed and weakened skin, as well as on varicose veins.

Enlarged prostate Following several clinical studies, nettle root is now commonly used as a first-line treatment in Europe to treat symptoms of enlarged prostate such as poor urine flow, pain or

---

KEY INFORMATION

SAFETY	★ ★ ★ ★ ☆
TRADITIONAL USE	★ ★ ★ ★ ☆
RESEARCH	★ ★ ★ ☆ ☆

BEST TAKEN AS Leaf: Infusion ✓✓✓ Soup ✓✓ Tincture ✓
Root: Tincture ✓✓✓ Tablet ✓✓
DOSAGE Leaf and root: A *(see pp.44–45)*
OFTEN USED WITH Calendula (*Calendula officinalis*)
CAUTIONS Fresh plant will sting! Rarely, can cause allergic skin reactions. Avoid root in pregnancy. See also pp.42–51.

difficulty in passing urine, and urinary frequency. Nettle root can be taken on its own as a tincture, or combined with saw palmetto (*Serenoa repens*).

Other uses Taken internally, nettle helps to prevent or stop bleeding from wounds and nosebleeds; it is a valuable remedy for heavy menstrual bleeding, reducing blood loss and, given its appreciable iron content, helping to prevent anaemia. Nettle leaf has anti-allergenic activity and is a useful addition to formulations for hay fever and asthma. Nettle leaf infusion makes an effective hair rinse.

CAPSULES

Found in temperate regions worldwide, nettle has been used medicinally for centuries. The aerial parts are picked in summer, while the shoots are picked in spring and eaten as a tonic food and vegetable.

NETTLE AS A VEGETABLE

In spring or early summer, don a pair of gloves and collect fresh young nettles – shoots, stems, and leaves – from unpolluted, unsprayed areas. Remove thick or old stems. Wash thoroughly, place in a non-aluminium saucepan, cover and simmer for 5 minutes (no added water is needed). Add butter or margarine and salt and pepper, as required. Serve warm, puréed in a blender if required. Alternatively, use as a base for nettle soup.

NETTLE SOUP

BILBERRY, BLUEBERRY

Vaccinium spp.

Bilberry tones up small blood vessels, especially capillaries, that permeate the tissues of the body. In particular, bilberry acts on the micro-circulation of the eye and can help to improve night vision and eyesight.

MEDICINAL USES

Parts used Fruit • Juice • Leaf

Key actions Anti-inflammatory • Anti-oedema (prevents fluid retention) • Antioxidant • Astringent • Circulatory tonic (fruit) • Urinary antiseptic (leaf)

Eyesight aid Taken long term, bilberry improves eye health, protecting against damage to the eye resulting from diabetes and high blood pressure. It can sometimes help in short-sightedness, and in the prevention of cataract formation.

Other uses Bilberry helps to improve poor peripheral circulation and prevents fluid leakage from blood vessels. Many circulatory problems can benefit from taking bilberry, including haemorrhoids and varicose veins, chilblains, Raynaud's disease, intermittent claudication, and easy bruising. Bilberries can be taken to relieve diarrhoea or constipation, and the leaves are a useful urinary antiseptic for conditions such as cystitis.

KEY INFORMATION

SAFETY	★ ★ ★ ★ ⯪
TRADITIONAL USE	★ ★ ★ ★ ☆
RESEARCH	★ ★ ★ ⯪ ☆

BEST TAKEN AS Fruit ✓✓✓ Tincture ✓✓ Tablet ✓

DOSAGE A (*see pp.44–45*)

OFTEN USED WITH Hawthorn (*Crataegus* spp.)

CAUTIONS Very high doses may increase the effect of prescribed anticoagulant medication. See also pp. 42–51.

DRIED LEAVES

berries tone the capillaries

DRIED BERRIES

Bilberry's potent antioxidant activity makes it a useful supplement in many chronic health problems, especially where the circulation is poor.

VALERIAN

Valeriana officinalis

Used wherever nervous tension, overactivity, or an inability to relax are present, valerian's gently sedative action helps to soothe and slow a nervous system that is beginning to spin out of control. It is one of the first herbs to consider when a remedy is needed to ease anxiety and a sense of panic.

MEDICINAL USES

Part used Root

Key actions Antispasmodic ● Mild analgesic ● Mild bitter ● Tranquilizer

Anxiety, nervous tension Safe and non-addictive, valerian helps to relieve anxiety, stress, and tension along with the familiar symptoms that go with them – tension headache, tensed muscles, eye strain, and nervous palpitations. Its tranquilizing action means that it can help to promote relaxation in chronic anxiety states, and it has been used to calm and improve sleep in children with autism and ADHD. People vary in their response to valerian – start with a low dose and build up. Valerian is usually best taken in small, frequent doses throughout the day.

KEY INFORMATION

SAFETY	★ ★ ★ ★ ☆
TRADITIONAL USE	★ ★ ★ ★ ☆
RESEARCH	★ ★ ★ ★ ☆

BEST TAKEN AS Tablet, capsule ✓✓✓ Tincture ✓✓
DOSAGE B *(see pp.44–45)*
OFTEN USED WITH Skullcap *(Scutellaria laterifolia)*
CAUTIONS Can cause drowsiness, for example, when driving or using machinery. Rarely, can cause headache or gastrointestinal upset, or worsen anxiety/insomnia. See also pp.42–51.

FRESH ROOT

Poor sleep

A key remedy in many herbal sleep preparations, valerian can prove valuable when sleep is disturbed due to worry or overwork. Combined with St John's wort (*Hypericum perforatum*), it improves sleep quality and eases anxiety and depression.

Other uses A good antispasmodic, valerian can relieve muscle pain and tension in menstrual cramps, rheumatic aches, and irritable bowel syndrome.

root calms anxiety

DRIED ROOT

Native to Europe and northern Asia, valerian grows in the wild in damp conditions. Its name is thought to be derived from the Latin *valere*, which means "to be well".

Primarily used as a stress-reliever, valerian (*Valeriana officinalis*) soothes nervous tension and anxiety. Its antispasmodic action works well in muscle pains and digestive cramps.

VERVAIN

Verbena officinalis

Used in Western and Chinese herbal traditions, vervain is restorative, acting mainly on the nervous and digestive systems. Traditional indications include nervous exhaustion, headaches, migraine, menstrual problems, weak digestive function, and urinary tract infection.

MEDICINAL USES

Parts used Aerial parts

Key actions Mild antidepressant
● Mild digestive tonic ● Nerve tonic
● Relaxant

Anxiety and nervous tension Thought to improve nervous vitality, vervain can be taken where long-term stress and worry are leading to nervous exhaustion. It is a useful remedy for migraine and stress-induced headaches. For best results, it should be taken for some weeks.

Premenstrual problems Thought to have mild progesterogenic activity, vervain is a valuable remedy for premenstrual tension and menstrual headaches, especially when combined with chaste berry (*Vitex agnus-castus*).

Other uses Vervain is an excellent remedy for poor appetite, especially where emotional factors are responsible. Useful therefore in anorexia nervosa, the herb is

tincture stimulates appetite

TINCTURE

best taken as a tincture before meals – if necessary, as drops in water or fruit juice. In China, vervain has been used to treat fever, such as in malaria. Although poorly researched, it appears to protect the liver. Vervain tea is taken traditionally to aid breast milk production.

A poorly researched remedy, vervain is prized by herbalists for its ability to restore a depleted nervous system and allay anxiety.

KEY INFORMATION

SAFETY ★★★★☆
TRADITIONAL USE ★★★★☆
RESEARCH ★☆☆☆☆
BEST TAKEN AS Infusion ✓✓✓ Tincture ✓✓
DOSAGE B (*see pp.44–45*)
OFTEN USED WITH St John's wort (*Hypericum perforatum*)
CAUTIONS Rarely, may cause skin rash.
See also pp.42-51.

CRAMP BARK, GUELDER ROSE

Viburnum opulus

Useful in problems affecting both skeletal muscle and internal organs, cramp bark lives up to its reputation as an effective antispasmodic. A key remedy in Western herbal medicine, cramp bark relaxes excessive muscle tone, thereby easing tensed and cramping muscles.

MEDICINAL USES

Part used Bark

Key actions Antispasmodic
● Astringent ● Lowers blood pressure

Muscle cramps and pains Rheumatic pain sometimes results more from locked muscles than inflammation. Here, cramp bark can prove particularly effective, relaxing tensed muscles and opening up the circulation to clear accumulated toxins – often a key factor in pain development. As well as relieving rheumatic and arthritic problems, cramp bark's antispasmodic action makes it a worthwhile treatment for restless legs, leg cramps, and spasmodic period pains. Take on its own to provide symptomatic relief, or combine with anti-inflammatory remedies and circulatory stimulants such as willow bark (*Salix alba*) and prickly ash (*Zanthoxylum* spp.) for rheumatic problems such as fibromyalgia. For period pains, take the remedy before pain begins.

Digestive cramps Cramp bark is effective for intestinal spasms, including irritable bowel syndrome.

Found growing in hedges, thickets, and woodland, cramp bark has distinctive red berries. It is native to Europe and the eastern regions of North America.

It combines well with chamomile (*Chamomilla recutita*) for cramps throughout the gastrointestinal tract.

Other uses Cramp bark is often included in formulations for high blood pressure, especially where tensed muscles are a feature.

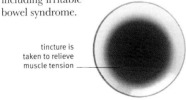

tincture is taken to relieve muscle tension

TINCTURE

DRIED BARK

KEY INFORMATION	
SAFETY	★ ★ ★ ★ ★
TRADITIONAL USE	★ ★ ★ ★ ☆
RESEARCH	★ ★ ☆ ☆ ☆
BEST TAKEN AS	Decoction ✓✓✓
	Tincture ✓✓ Capsule ✓
DOSAGE	C *(see pp.44–45)*
OFTEN USED WITH	Valerian
	(*Valeriana officinalis*)
CAUTIONS	None known. See also
	pp.42–51.

CHASTE BERRY

Vitex agnus-castus

One of the few herbal medicines known to have a progesterone-type activity within the body, chaste berry is a specific for menstrual and peri-menopausal problems. Acting on the pituitary gland at the base of the brain, it improves menstrual regularity.

MEDICINAL USES

Part used Fruit

Key actions Hormone balancer
● Stimulates breast milk

Menstrual disorders The essential natural remedy to try in the case of menstrual problems, chaste berry is a specific for menstrual irregularity and premenstrual syndrome. Though not suitable for all types of menstrual disturbance, it will often help to relieve menstrual symptoms such as breast tenderness, fluid retention, headache, and premenstrual tension. If it is taken over several months, such symptoms will usually become milder and of shorter duration. Tincture or extract is usually taken on rising in the morning, when the pituitary gland is most active. Chaste berry can be taken to treat heavy menstrual bleeding and period pains, but will work better when combined with other appropriate remedies prescribed by a qualified herbal practitioner.

Polycystic ovary disease Herbal medicine can be very helpful in controlling or

berries have progesterone activity

STEM WITH BERRIES

KEY INFORMATION

SAFETY ★★★★⯪
TRADITIONAL USE ★★★★★
RESEARCH ★★★★☆
BEST TAKEN AS Tincture ✓✓✓ Tablet ✓✓ Capsule ✓
DOSAGE M *(see pp.44–45)*. Take before breakfast each morning
OFTEN USED WITH Black cohosh (*Cimicifuga racemosa*)
CAUTIONS Concurrent use with contraceptive pill, fertility treatment, or hormone replacement therapy is not advisable. Avoid in pregnancy. Rarely, may cause gastrointestinal upset, headache, or dizziness. See also pp.42–51.

tincture is taken for irregular periods

TINCTURE

reversing this difficult problem. Though best treated professionally, self-treatment with chaste berry alone can sometimes bring about a significant improvement in symptoms. It may need to be taken for at least 3–4 months before results are seen.

Infertility Thought to have a pronounced ability to fine tune oestrogen and progesterone release through the menstrual cycle, chaste berry can improve fertility and increase the chances of conception. It is most likely to help where there are no structural factors involved.

Menopausal problems Most likely to be of value in the year or so before the menopause,

berries act on the
pituitary gland

DRIED BERRIES

chaste berry can help
to maintain a regular
menstrual cycle and control bleeding.
It may also be taken, typically with
remedies such as black cohosh
(*Cimicifuga racemosa*) and sage (*Salvia
officinalis*), to relieve or prevent
menopausal symptoms such as
headache, hot flushes, and night sweats.

Other uses Chaste berry is also used to
treat acne, which is often linked to raised
male hormone levels. Other hormonal
effects include increased breast milk
production in lactating mothers.

MONK'S PEPPER

In medieval times, monks reputedly chewed
chaste berries to curb their sexual desire. In a
similar vein, the 16th-century English herbalist
Gerard wrote: "Agnus castus (or chaste berry) is
a singular medicine for such as would willingly
live chaste, for it withstandeth all uncleanness
or desire to the flesh,
consuming or drying
up the seed of
generation, in what
sort soever it bee
taken ... for which
cause it was called
castus, that is to
say chaste, cleane,
and pure."

Chaste berry has a long history of use: Homer
referred to it in his epic poem the *Iliad* as a herb
employed to keep all evils at bay.

HEARTSEASE, WILD PANSY

Viola tricolor

Perhaps appreciated more for its beautiful flowers than as a medicine, heartsease is a valued traditional remedy for skin and chest problems.

MEDICINAL USES

Parts used Aerial parts

Key actions Anti-inflammatory
● Diuretic ● Expectorant

Skin and chest problems Heartease is often combined with herbs such as nettle (*Urtica dioica*) and red clover (*Trifolium pratense*) for skin disorders and to aid detoxification. Commonly used for eczema and other itchy skin conditions, especially in children, heartsease can be taken internally as a tincture or infusion; the latter can be applied to itchy sites. For chesty coughs and bronchitis, it combines well with thyme (*Thymus vulgaris*).

FLOWERS AND FOLIAGE

KEY INFORMATION

SAFETY	★ ★ ★ ★ ☆
TRADITIONAL USE	★ ★ ★ ☆ ☆
RESEARCH	★ ✬ ☆ ☆ ☆
BEST TAKEN AS	Infusion ✓✓✓ Tincture ✓✓
DOSAGE	B (*see pp.44–45*)
CAUTIONS	None known. See also pp.42–51.

MISTLETOE

Viscum album

The main therapeutic value of mistletoe lies in treating high blood pressure, though it has traditionally been used to treat epilepsy, insomnia, and tinnitus.

MEDICINAL USES

Parts used Aerial parts (not berries)

Key actions Lowers blood pressure
● Mild sedative

Cardiovascular problems Mildly sedative, mistletoe helps to reduce over-activity, relax blood vessels, and prevent panic attacks. Traditionally thought to act as a heart tonic, mistletoe can prove a useful addition to a formulation for high blood pressure when taken orally. It is best taken on professional advice.

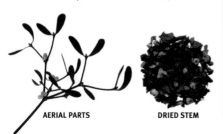

AERIAL PARTS   DRIED STEM

KEY INFORMATION

SAFETY	★ ★ ★ ✬ ☆
TRADITIONAL USE	★ ★ ★ ✬ ☆
RESEARCH	★ ★ ☆ ☆ ☆
BEST TAKEN AS	Tincture ✓✓✓ Capsule ✓✓ Tablet ✓
DOSAGE	M, C (*see pp.44–45*)
CAUTIONS	Avoid in pregnancy. Potentially toxic at high dosage. See also pp.42–51.

GRAPE VINE

Vitis vinifera

Grapes have astringent, laxative, and tonic qualities, and are almost universally recommended for convalescence – flowers and grapes are the gifts one brings to the ill and infirm. Traditionally, grapes are used to cool fevers and, as part of a grape fast, promote tissue cleansing.

MEDICINAL USES

Parts used Fresh or dried fruit ● Leaf ● Seed ● Seed oil

Key actions Antioxidant ● Nutritive ● Tonic

Circulatory tonic The beneficial effects of red wine on the heart and circulation are well known, though red grape juice may be as good if not better. Research has confirmed the antioxidant properties of the red pigments in red grapes. Similar in many respects to the antioxidants in bilberry (*Vaccinium myrtillus*) and maritime pine (*Pinus maritima*), grape seed extract provides powerful antioxidant support to tissues under stress, increasing vitamin C levels within the cells and strengthening blood vessels, particularly small arteries. It is a valuable supplement in chronic conditions affecting the circulation, notably furring up of the arteries (atherosclerosis), peripheral vascular disease, including easy bruising, varicose veins, and peripheral neuropathy associated with diabetes.

Sluggish liver and kidneys A grape fast, a naturopathic cleansing regime in which one eats only grapes for several days, helps in detoxifying the body, especially in serious ill health. Although

Native to southern Europe and western Asia, the grape vine is cultivated in warm temperate regions throughout the world for its fruit and to produce wine.

not suitable for everyone, a grape fast can improve health and vitality where liver and kidneys are sluggish. Follow only on professional advice.

Other uses The seeds and leaves are astringent and anti-inflammatory, and have been taken to relieve diarrhoea. Raisins are highly nutritious, gently laxative, and demulcent.

KEY INFORMATION

SAFETY	★ ★ ★ ★ ★
TRADITIONAL USE	★ ★ ★ ⯪ ☆
RESEARCH	★ ★ ★ ⯪ ☆

BEST TAKEN AS Fruit ✓✓✓ Seed extract ✓✓✓ Juice ✓✓

DOSAGE Food; M (*see pp.44–45*)

CAUTIONS None known. See also pp.42–51.

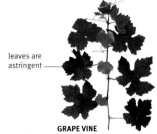

leaves are astringent

GRAPE VINE

English herbalist Nicholas Culpeper referred to grape vine (*Vitis vinifera*) as "a most gallant tree of the sun very sympathetical to the body of man." Antioxidant and tonic, it has many health benefits.

WITHANIA, ASHWAGANDHA

Withania somnifera

An increasingly talked about herb in the West, withania has been prized in Ayurvedic medicine for over 4,000 years. Often compared to ginseng, withania is more a sedative than a stimulant, its calming, restorative action helping to relieve stress and exhaustion.

MEDICINAL USES

Parts used Root • Fruit

Key actions Adaptogen • Anti-inflammatory • Sedative • Tonic

Exhaustion and nervous debility

Withania is a first-rate tonic and building herb, especially useful in conditions involving chronic weakness and nervous debility. It is a good remedy to quieten anxiety states and overactivity and makes an excellent restorative for old age and convalescence. Root and fruit have been used traditionally as a remedy for senile dementia. For best results, the herb should be taken for several months.

Weak immune system Taken long term, withania strengthens connective tissue and supports balanced immune function, leading to increased vigour and raised white blood cell count. It therefore has a role to play in a wide range of chronic illnesses, particularly those involving chronic inflammation such as fibromyalgia and psoriasis. In such conditions, it is best taken on professional advice.

Other uses Traditionally valued for its aphrodisiac properties, withania can be taken to improve erectile dysfunction and enhance fertility in men and women. The root and fruit are used in Ayurveda to treat respiratory conditions such as asthma and bronchitis.

KEY INFORMATION	
SAFETY	★★★★✩
TRADITIONAL USE	★★★★★
RESEARCH	★★★★☆
BEST TAKEN AS	Tablet ✓✓✓ Tincture ✓✓ Decoction ✓
DOSAGE	B (*see pp.44–45*)
OFTEN USED WITH	Siberian ginseng (*Eleutherococcus senticosus*)
CAUTIONS	Avoid during pregnancy. See also pp.42–51.

berries can be taken
in convalescence

DRIED BERRIES

root is useful
in anaemia

FRESH ROOT

This herb is known as "ashwagandha" in Sanskrit, meaning "horse's smell". It also implies a horse's strength, indicative of its use as a strengthening herb.

PRICKLY ASH

Zanthoxylum clava-herculis

A highly valued North American remedy, prickly ash has traditionally been used for numerous conditions, ranging from toothache and rheumatic pain to cramps and poor peripheral circulation. Although bitter and hot to taste, the bark has long been a stand by method for cleaning the teeth.

MEDICINAL USES

Parts used Bark ● Fruit

Key actions Analgesic ● Anti-rheumatic ● Circulatory stimulant ● Stimulates sweating

Rheumatic and arthritic problems
By promoting local blood flow and the clearance of waste products, prickly

Prickly ash bark has a stronger stimulant action on blood flow than the berries, and is normally used when treating peripheral circulatory disorders.

FRESH BARK

ash can bring relief wherever muscle tension or poor circulation have led to the development of rheumatic symptoms. It can be particularly helpful in relieving chronic musculoskeletal problems such as fibromyalgia. It is generally best combined with anti-inflammatory or anti-rheumatic remedies such as meadowsweet (*Filipendula ulmaria*).

Poor peripheral circulation One of the best remedies for weak circulation, prickly ash is thought to stimulate arterial blood flow. Taken over several weeks or months together with remedies such as cramp bark (*Viburnum opulus*), it can significantly improve peripheral blood flow to the hands and feet. Conditions such as carpal tunnel syndrome, intermittent claudication, Raynaud's disease, varicose veins, and haemorrhoids can all benefit.

DRIED BERRIES

KEY INFORMATION

SAFETY	★ ★ ★ ★ ☆
TRADITIONAL USE	★ ★ ★ ★ ☆
RESEARCH	★ ☆ ☆ ☆ ☆

BEST TAKEN AS Tincture ✓✓✓ Capsule ✓✓ Tablet ✓

DOSAGE C *(see pp.44–45)*

OFTEN USED WITH Willow bark (*Salix alba*)

CAUTIONS Avoid in pregnancy and while breast-feeding. See also pp.42–51.

CORNSILK

Zea mays

Maize is one of the world's most popular foods, yet few realize that the silky brown fronds wrapped around the cob make a valuable medicine. Best prepared as an infusion, cornsilk has an affinity for the urinary system, soothing and protecting the kidneys, bladder, and urinary tract.

MEDICINAL USES

Part used "Silk" or fronds (pistils)

Key actions Demulcent ● Diuretic
● Mild urinary antiseptic
● Wound healer

Urinary tract problems With diuretic, demulcent, and mild antiseptic activity, cornsilk is a remedy to take at the first sign of urinary infection, helping to soothe inflammation and irritation and flush out infection. Protective and restorative rather than a front line treatment for infection, cornsilk supports kidney function and the health of the urinary tract. While it is not an effective treatment on its own, take the infusion to aid recovery from cystitis and as a preventative against recurring infection. The infusion may also help to ease bladder irritability and poor urine flow. Use 1 dessertspoonful of chopped cornsilk to a cup and brew for 15 minutes. Drink up to 5 cups a day or

DRIED CORNSILK

as required. Other urinary-related problems such as chronic urethritis and an enlarged prostate can benefit from this gentle-acting herb. In cases of kidney disease, including kidney stones, take only on professional advice.

Other uses Other indications for cornsilk include high blood pressure and fluid retention. Despite its diuretic activity, cornsilk is worth trying in problems such as stress or pressure incontinence and bed-wetting.

Ground cornflour mixed to a paste with a little water makes a good poultice for drawing out a stubborn and painful splinter.

KEY INFORMATION

SAFETY	★ ★ ★ ★ ☆
TRADITIONAL USE	★ ★ ★ ★ ☆
RESEARCH	★ ★ ☆ ☆ ☆

BEST TAKEN AS Infusion ✓✓✓
Tincture ✓✓
DOSAGE A (*see pp.44–45*)
OFTEN USED WITH Cranberry (*Vaccinium macrocarpon*)
CAUTIONS None known at normal dosage. See also pp.42–51.

Cultivated for more than 4,000 years in Mexico as a food crop, sweetcorn (*Zea mays*) and the silky fronds that surround it are traditionally used to treat kidney and urinary tract disorders.

Ginger is native to South East Asia. More than 50 per cent of the world's ginger is still grown in India, Nepal, and China.

GINGER

Zingiber officinale

The warm taste of ginger, one of the most versatile of all spices, adds zest to any herbal infusion. Taken on its own, it stimulates circulation to the skin, promotes sweating, and relieves nausea. Combine the fresh root with garlic and honey to bring quick relief to colds and flu and settle stomach upset.

MEDICINAL USES

Part used Root (fresh and dry)

Key actions Anti-emetic
- Anti-inflammatory ● Antioxidant
- Circulatory stimulant
- Digestive tonic ● Stimulates sweating

Poor circulation Whether taken on its own or combined with a circulatory stimulant such as ginkgo (*Ginkgo biloba*), ginger helps to tone capillaries and will benefit any condition involving weak or deficient circulation, especially to the

root has anti-inflammatory activity

SLICED ROOT

head or limbs. Regular intake of ginger – as an infusion, tincture, or capsule – can make an appreciable difference where poor peripheral circulation is linked to weak digestive function. Ginger has blood-thinning properties, so should not be taken at above 2g dried (4g fresh) root by those taking anticoagulants.

Nausea and vomiting Probably ginger's most valuable area of activity, supported by in-depth research, is as a safe and effective remedy for nausea, vomiting, motion sickness, and morning sickness during pregnancy. Take ginger tea or a

Pungent and somewhat lemony in taste, ginger has a multitude of uses that have gained it the label of "the best medicine in the world".

KEY INFORMATION

SAFETY	★ ★ ★ ★ ✯
TRADITIONAL USE	★ ★ ★ ★ ★
RESEARCH	★ ★ ★ ✯ ☆

BEST TAKEN AS Infusion ✓✓✓
Capsule ✓✓ Tincture ✓
DOSAGE Fresh root: C ; Dried root: D
(see pp.44–45)
OFTEN USED WITH Garlic (*Allium sativum*)
CAUTIONS Maximum dose in pregnancy and if taking anticoagulants is 2g dried (4g) fresh root a day. Can cause discomfort or burning in stomach disorders. See also pp.42–51.

standardized extract at the earliest signs of symptoms. If using for travel sickness, start taking the tea or extract before commencing your journey. Ginger also makes a good remedy for symptoms such as intestinal colic, wind, and bloating. An under-appreciated remedy for gastro-intestinal infection, it can provide significant relief in symptoms such as indigestion, bloating, and diarrhoea.

Anti-inflammatory action Recent research has shown ginger root to have potent anti-inflammatory activity, making it a possible alternative to aspirin-type medicines in treating arthritic pain. As relatively large doses are required, seek professional advice in this situation.

Viral infection When a cold, flu, cough, or chest infection threatens, fresh ginger root tea can improve resistance as well as one's sense of well-being. The tea combines well with other remedies such as cinnamon, garlic, and liquorice.

DIGESTIVE REMEDY

Ginger's positive effects on the digestive system are the subject of ongoing scientific study. Several clinical trials have shown that ginger extract relieves post-operative nausea and vomiting, with a low incidence of side effects. Evidence also supports ginger's use in relieving travel sickness, as well as nausea and vomiting during pregnancy. Active constituents in ginger are thought to stimulate stomach activity and to relieve spasm, which also helps in gastrointestinal disorders such as cramps, colic, and diarrhoea.

CAPSULES

Other uses Ginger is included in numerous herbal formulations and can, where indicated, be combined with almost any other remedy. Some of the conditions that it can benefit include: period pain (take symptomatically with remedies such as cramp bark (*Viburnum opulus*); irregular menstrual cycle; anaemia and lowered vitality, where it combines well with withania (*Withania somnifera*); and headache and migraine. Evidence also points to ginger root lowering cholesterol levels and protecting against stomach ulcers.

FRESH SHOOT

root stimulates peripheral circulation

POWDERED ROOT

root has anti-viral properties

FRESH ROOT

COMMON HEALTH PROBLEMS

REMEDIES FOR HOME USE

The following section gives straightforward recommendations for remedies that can be safely used at home to treat many common health problems. A number of remedies are suggested for each condition, and these can be used individually or in combination.

How to use this section

GENERAL GUIDELINES

1. Be clear about what condition it is that needs treating. If you are unsure, seek professional advice, such as a telephone help line.
2. Select the herb(s) you wish to use from the list and look up in A–Z of Herbal Remedies, pp.52–255. Note the dosage and cautions listed in the key information box (see also p.45).
3. Decide how to take the herb(s), for example as an infusion.
4. Work out an appropriate dosage (see also pp.44–45): Take a single remedy as recommended; for combined remedies, work out which herb has the lowest recommended dosage and take the combination at this dosage. Medium to high doses can be taken for up to 4 days; lower doses are required when taking a herbal remedy longer term.
5. Teas and decoctions: the dosages given apply when making teas and decoctions from dried herb material – bark, leaves, roots, etc. For fresh herb material you can use $1^{1}/_{2}$–2 times the quantity of dried material.
6. Tinctures: It is not possible to give clear guidelines for tinctures owing to the wide variation in their strength. Ask advice on dosage when purchasing a tincture. In general, the dosage range for a 1:3 tincture is the same (in millilitres not grams) as the dosages in the Recommended adult dosage box (left), i.e. for A, the dosage of a 1:3 tincture is 5–15ml a day.
7. Powders: take the minimum recommended daily dosage only.
8. Tablets and capsules: Take at the manufacturer's recommended dosage.

RECOMMENDED ADULT DOSAGE

See also pp.44–45. If over 70, see p.45. For children, see pp.44, 47, and 273–275. Look up the dosage for each remedy.

A	5–15g a day, or max. 100g ($3^{1}/_{2}$ oz) per week
B	3–7.5g a day, or max. 50g (2 oz) per week
C	2–4g a day, or max. 30g (1 oz) per week
D	1–2g a day, or max 15g ($^{1}/_{2}$ oz) per week
M	Take product at manufacturer's recommended dosage
T	Topical application on the skin only (not to be taken internally)

KEY TO PAGES 259–275

T	TEA OR INFUSION (pp.31, 34)
D	DECOCTION (pp.31, 34)
Tr	TINCTURE (pp.32, 35)
C	CAPSULE OR TABLET (p.32)
S	SYRUP (p.32)
G	GARGLE
EO	ESSENTIAL OIL, external use only (p.33)
L	LOCAL USE, cream, ointment, lotion; other application in brackets (p.33)

HEAD

CONDITION	HERB	PREPARATION
Tension headache	Limeflower (pp.222–223)	T, Tr, C
	Lavender (pp.152–153)	L (EO)
	Skullcap (p.206)	T, Tr, C
	Guarana (p.172)	T, C
Migraine	Feverfew (p.214)	Tr, C
	Butterbur (p.178)	C
	Lavender (pp.152–153)	L (EO)
Menstrual headache	Chaste berry (pp.242–243)	Tr, C
	Skullcap (p.206)	T, Tr, C
	Feverfew (p.214)	Tr, C
Dizziness	Black cohosh (pp.100–101)	Tr
	Ginkgo (pp.138–139)	Tr, C
	Rosemary (p.190)	T, Tr, C
Tinnitus	Ginkgo (pp.138–139)	Tr, C
	Mistletoe (p.244)	Tr, C
	Black cohosh (pp.100–101)	Tr, C
Earache	Garlic (pp.60–61)	L, oil
	Elderflower (pp.204–205)	T, Tr, C, S
	Golden seal (p.147)	Tr, C
	Lavender (pp.152–153)	L (EO)
Infection of ears, sinuses, and nose	Elderberry (pp.204–205)	S Tr, C
	Echinacea (pp.118–119)	T, Tr, C
	Chiretta (p.55)	Tr, C
	Thyme (p.221)	T, Tr, C
Catarrh and congestion	Elderflower (pp.204–205)	T, Tr, C
	Plantain (p.181)	T, Tr, C
	Golden seal (p.147)	Tr, C
	Sage (p.201)	T, Tr, C
Hay fever and allergy	Eyebright (p.129)	T, Tr, C
	Baical skullcap (p.203)	T, Tr, C
	Butterbur (p.178)	C
	Elderflower (pp.204–205)	T, Tr, C
Nosebleed	Yarrow (p.54)	T, Tr, C
	Nettle (pp.234–235)	T, Tr, C
	Golden seal (p.147)	Tr, C

HEAD (CONTINUED)

CONDITION	HERB	PREPARATION
Styes	Calendula (p.86)	L (ointment)
	Chamomile (pp.98–99)	L (lotion)
	Golden seal (p.147)	L (ointment)
Conjunctivitis	Eyebright (p.129)	L (lotion) T, Tr, C
	Echinacea (pp.118–119)	T, Tr, C
	Witch hazel (p.144)	L (distilled water)
Sore eyes/lids	Witch hazel (p.144)	L (distilled water)
	Chamomile (pp.98–99)	L (T)
Poor eyesight	Bilberry (p.236)	T, Tr, C S
	Ginkgo (pp.138–139)	Tr, C
	Eyebright (p.129)	T, Tr, C
Toothache	Clove (pp.37, 123)	L (1 clove or 1 drop EO)
Dental treatment	St John's wort (pp.148–149)	L (oil)
Mouth ulcers	Myrrh (p.107)	L (diluted Tr)
	Liquorice (pp.140–141)	L (neat Tr)
	Echinacea (pp.118–119)	L (diluted Tr)
Gum problems	Bilberry (p.236)	L (T or Tr)
	Myrrh (p.107)	L (diluted Tr)
	Yarrow (p.54)	L (T or Tr)
Thrush (mouth)	Cat's claw (p.232)	Tr, C, L (diluted Tr)
	Pau d'arco (p.215)	Tr, C, L (diluted Tr)

CAUTIONS FOR HEAD CONDITIONS

Seek immediate professional advice for:
- Fever of 39°C (102°F) or above
- Heavy nosebleed lasting more than 1 hour
- Persistent one-sided headache
- Headache or pain that fails to improve within 48 hours despite self-medication
- Double vision/visual disturbance
- Unexplained dizziness
- Sudden or severe allergy

THROAT, CHEST, AND LUNGS

CONDITION	HERB	PREPARATION
Colds, flu colds	Ginger (pp.254–255)	T, Tr, C
	Chiretta (p.55)	Tr, C
	Echinacea (pp.118–119)	T, Tr, C
	Cinnamon (p.97)	T, Tr, C
	Yarrow (p.54)	T, Tr
	Elderflower/berry (pp.204–205)	T, Tr, S
Sore throat/hoarseness	Sage (p.201)	G T, Tr
	Echinacea (pp.118–119)	G T, Tr
	Plantain (p.181)	G T, Tr
Catarrh and congestion	Elderflower (pp.204–205)	T, Tr, C
	Thyme (p.221)	T, Tr, C
	Elecampane (p.150)	T, Tr, C
	Garlic (pp.60–61)	C, or with food
Cough	Thyme (p.221)	T, Tr, C
	Liquorice (pp.140–141)	T, Tr, C
	Elecampane (p.150)	T, Tr, C
	Caraway (p.94)	T, Tr, C
Shortness of breath	Baical skullcap (p.203)	T, Tr, C
	Angelica (p.67)	T, Tr, C
	Thyme (p.221)	T, Tr, C
	Lobelia (p.157)	C
Bronchitis	Elecampane (p.150)	T, Tr, C
	Thyme (p.221)	T, Tr, C
	Caraway (p.94)	T, Tr, C
	Lobelia (p.157)	C
Preventing infection	Elderberry (p.204)	T, Tr, S C
	Sea buckthorn (p.146)	T, S, C
	Thyme (p.221)	T, Tr, C
	Echinacea (pp.118–119)	T, Tr, C

CAUTIONS FOR THROAT, CHEST, AND LUNG CONDITIONS

Seek immediate professional advice for:
- Fever of 39°C (102°F) or above
- Allergic reactions, including asthma
- Allergies that worsen after taking herbal remedies
- Persistent hoarseness, cough, or sore throat
- Chest pain or shortness of breath
- Coughing up blood

DIGESTION

CONDITION	HERB	PREPARATION
Poor appetite	Gentian (p.136)	Tr,
	Alfalfa (p.162)	Tr, C
	Angelica (p.67)	Tr
Acid indigestion/reflux	Meadowsweet (p.130)	T, Tr
	Marshmallow (p.66)	T, Tr
	Liquorice (pp.140–141)	T, Tr
	Chamomile (pp.98–99)	T, Tr
Nausea/motion sickness	Ginger (pp.254–255)	T, Tr, C
	Chamomile (pp.98–99)	T, Tr, C
	Lemon balm (p.164)	T, Tr,
Indigestion, wind, bloating	Fennel (p.131)	T, Tr, C
	Cardamon (p.122)	T, Tr, C
	Peppermint (p.165)	T, Tr, C
	Oregano (p.171)	T, Tr, C
Digestive infections	Garlic (pp.60–61)	C, or with food
	Cat's claw (p.232)	Tr, C
	Golden seal (p.147)	Tr, C
	Calendula (p.86)	T, Tr, C
	Cinnamon (p.97)	T, Tr, C
Worms	Pumpkin seeds (p.109)	Ground seeds
	Thyme (p.221)	T, Tr, C
	Garlic (pp.60–61)	C, or with food
Cramps and pain	Cramp bark (p.241)	T, Tr, C
	Chamomile (pp.98–99)	T, Tr, C
	Ginger (pp.254–255)	T, Tr, C
Diarrhoea	Plantain (p.181)	T, Tr, C
	Slippery elm (p.231)	T, C
	Raspberry leaf (p.191)	T, Tr, C
Constipation	Flaxseed (p.158–159)	T
	Chinese rhubarb (p.185)	Tr, C
	Senna (p.87)	T, C
Irritable bowel	Peppermint (p.165)	T, Tr, C
	Valerian (p.237)	T, Tr, C
	Clove (p.123)	T, Tr, C
	Aloe vera (pp.62–63)	Juice or C

DIGESTION (CONTINUED)

CONDITION	HERB	PREPARATION
Supporting liver/ gall-bladder	Milk thistle (p.208)	C
	Dandelion (p.216)	T, Tr, C
	Bupleurum (p.83)	C Tr
	Schisandra (p.202)	T, Tr, C
Helping weight loss	Kelp (p.134)	C Tr
	Globe artichoke (p.113)	C Tr
Helping weight gain	Alfalfa (p.162)	T, or as food
	Fenugreek (p.225)	T, C, or as food

CAUTIONS FOR DIGESTIVE CONDITIONS

Seek immediate professional advice for:
• Difficulty in swallowing
• Persistent abdominal pain or indigestion
• Change in bowel habit
• Passing blood in the stools
• Persistent weight loss

CIRCULATION AND HEART

CONDITION	HERB	PREPARATION
High blood pressure	Hawthorn (pp.110–111)	T, Tr, C
	Dan shen (p.197)	T, Tr, C
	Dandelion leaf (p.216)	T, Tr, C
	Yarrow (p.54)	T, Tr, C
Low blood pressure	Rosemary (p.190)	T, Tr, C
	Liquorice (pp.140–141)	Tr, C
	Nettle leaf (pp.234–235)	T, Tr, C
Palpitations	Motherwort (p.156)	T, Tr, C
	Limeflower (pp.222–223)	T, Tr, C
	Lemon balm (p.164)	T, Tr, C
Poor peripheral circulation	Ginkgo (pp.138–139)	Tr, C
	Cinnamon (p.97)	T, Tr, C
	Cayenne pepper (pp.90–91)	C, or in food

CIRCULATION AND HEART (CONTINUED)

CONDITION	HERB	PREPARATION
Varicose veins/ haemorrhoids	Horse chestnut (pp.56–57)	C, Tr
	Butcher's broom (p.194)	C, Tr
	Witch hazel (p.144)	L (distilled water)
	Gotu kola (p.95)	Tr, C
	Bilberry (p.236)	T, Tr, C
Poor healing	Gotu kola (p.95)	Tr, C
	Yarrow (p.54)	T, Tr, C
	Plantain (p.181)	T, Tr, C
	Comfrey (pp.212)	L
Supporting heart and circulation	Hawthorn (pp.110–111)	C, Tr, T
	Motherwort (p.156)	T, Tr, C
	Garlic (pp.60–61)	C, or with food

CAUTIONS FOR CIRCULATION AND HEART CONDITIONS

Seek immediate professional advice for:
• Chest pain or shortness of breath
• Unexplained dizziness
• Hot, swollen, or ulcerated tender veins

BLOOD, METABOLIC

CONDITION	HERB	PREPARATION
Anaemia	Nettle (pp.234–235)	T, Tr, C
	Parsley leaf (p.179)	T, or as food
	Gentian (p.136)	Tr
Raised cholesterol levels	Guggul (p.107)	C
	Turmeric (p.112)	C
	Alfalfa (p.162)	T, or as food
	Globe artichoke (p.113)	C
Unstable blood sugar levels	Holy basil (p.167)	T, Tr, C
	Gymnema (p.136)	Tr, C
	Cinnamon (p.97)	T, Tr, C

BLOOD, METABOLIC (CONTINUED)

CONDITION	HERB	PREPARATION
Overactive thyroid	Motherwort (p.156)	T, Tr, C
	Lemon balm (p.164)	T, Tr, C
	Withania (p.248)	Tr, C
Underactive thyroid	Kelp (p.134)	C
	Cayenne pepper (pp.90–91)	C
	Siberian ginseng (p.124–125)	Tr, C

CAUTIONS FOR BLOOD, METABOLIC CONDITIONS

Seek immediate professional advice for:
• Persistent weight loss
• Frequent and excessive urination

BLADDER AND URINARY TRACT

CONDITION	HERB	PREPARATION
Cystitis	Cranberry (p.233)	Juice, Tr, C
	Buchu (p.76)	T, Tr, C
	Crataeva (p.108)	T, Tr, C
	Cornsilk (p.250)	T, Tr, C
Urethritis	Cranberry (p.233)	Juice, Tr, C
	Cornsilk (p.250)	T, Tr, C
	Puncture vine (p.224)	Tr, C
Frequency	Cornsilk (p.250)	T, Tr, C
	Cramp bark (p.241)	T, Tr, C
	Passion flower (p.173)	T, Tr, C

CAUTIONS FOR BLADDER AND URINARY TRACT CONDITIONS

Seek immediate professional advice for:
• Passing blood in the urine
• Pain in the kidneys
• Fever of 39°C (102°F) or above
• Urinary infections that deteriorate despite taking herbal remedies

JOINTS, MUSCLES, AND BONE

CONDITION	HERB	PREPARATION
Sprains, bruises, and sports injuries	Comfrey (p.212)	L
	Arnica (p.66)	L
	Gotu kola (p.95)	Tr, C
Fractures	Comfrey (p.212)	L
	Plantain (p.181)	L, Tr, C
	Yarrow (p.54)	L, Tr, C
Muscular aches and pains	Birch (p.81)	T, Tr, C
	Prickly ash (p.249)	Tr, C
	Bogbean (p.166)	Tr, C
	Meadowsweet (p.130)	T, Tr, C
Joint pain and stiffness	Devil's claw (p.145)	C, Tr
	Boswellia (p.82)	C
	Willow bark (p.196)	C, Tr
	Celery seed (p.69)	C, Tr
Chronic inflammation	Boswellia (p.82)	C
	Willow bark (p.196)	C, Tr
	Golden root (p.186)	C, Tr
	Turmeric (p.112)	C, Tr
Back problems	Cramp bark (p.241)	Tr, C
	St John's wort (pp.148–149)	T, Tr, C, L (oil)
	Prickly ash (p.249)	Tr, C
	Boswellia (p.82)	C
Restless legs	Cramp bark (p.241)	Tr, C
	Chamomile (pp.98–99)	T, Tr, C
	Prickly ash (p.249)	Tr, C
	Lavender (pp.152–153)	L (EO)
Supporting muscular-skeletal health as food	Alfalfa (p.162)	T, or as food
	Flaxseed (pp.158–159)	Ground seed
	Meadowsweet (p.130)	T, Tr, C

CAUTIONS FOR JOINT, MUSCLE, AND BONE CONDITIONS

Seek immediate professional advice for:
- Frequent and persistent back pain
- Unexplained leg pain and swelling
- Broken or suspected broken bones
- Any injury that may need an X-ray

SKIN

CONDITION	HERB	PREPARATION
Acne and boils	Burdock (p.72)	D, Tr, C
	Yellow dock (p.195)	D, Tr, C
	Red clover (pp.226–227)	T, Tr, C
	Echinacea (pp.118–119)	T, Tr, C
	Tea tree (p.163)	L (EO)
Nettle rash	Nettle leaf (pp.234–235)	T, Tr, C
	Baical skullcap (p.203)	T, Tr, C
	Chamomile (pp.98–99)	T, Tr, C
	Aloe vera (pp.62–63)	L (juice/gel)
Itchiness	Chickweed (p.213)	L
	Witch hazel (p.144)	L
	Evening primrose oil (p.170)	L
	Borage oil (p.80)	L
Eczema	Oregon grape (p.77)	D, Tr, C
	Calendula (p.86)	T, Tr, C
	Burdock (p.72)	D, Tr, C
	Gotu kola (p.95)	T, Tr, C
Fungal infections	Echinacea (pp.118–119)	Tr, C
	Pau d'arco (p.215)	D, Tr, C
	Thyme (p.221)	T, Tr, C
	Tea tree (p.163)	L (EO)
Herpes sores/shingles	Echinacea (pp.118–119)	Tr, C
	Lemon balm (p.164)	T, Tr, C
	St John's wort (pp.148–149)	T, Tr, C
	Pau d'arco (p.215)	D, Tr, C
Warts	Thuja (p.220)	Tr
	Garlic (pp.60–61)	L
Cuts, grazes, and minor wounds	Calendula (p.86)	L
	Aloe vera (pp.62–63)	Juice/gel
	Yarrow (p.54)	L
Bruises	Arnica (p.66)	L
	Comfrey (p.212)	L
	Plantain (p.181)	T, Tr, C, L
	Gotu kola (p.95)	T, Tr, C, L

SKIN

CONDITION	HERB	PREPARATION
Minor burns	Aloe vera (pp.62–63)	Juice/gel
	Calendula (p.86)	L (lotion or cream)
	St John's wort (pp.148–149)	L (oil)
	Witch hazel (p.144)	L (distilled water)
Sunburn	Aloe vera (pp.62–63)	Juice/gel
	Calendula (p.86)	L
Headlice	Neem (p.75)	L
Detox/supporting skin health	Dandelion root (p.216)	D, Tr, C
	Nettle (pp.234–235)	T, Tr, C
	Red clover (pp.226–227)	T, Tr, C
	Yellow dock (p.195)	D, Tr, C

CAUTIONS FOR SKIN PROBLEMS

Seek immediate professional advice for:
- Serious wounds, grazes, bruising, bites, and stings
- Sudden swelling or allergic reaction
- Non-minor burns, including sunburn
- A mole that has changed shape, size, or colour or itches or bleeds
- A sore or boil that does not heal, or unexplained swellings under the skin
- Shingles or suspected shingles

WOMEN'S HEALTH PROBLEMS

CONDITION	HERB	PREPARATION
Premenstrual tension	Chaste berry (pp.242–243)	Tr, C
	Black cohosh (pp.100–101)	Tr, C
	Skullcap (p.206)	T, Tr, C
	Evening primrose oil (p.170)	C
Pain/cramps	Cramp bark (p.241)	T, Tr, C
	White peony (p.172)	T, Tr, C
	Skullcap (p.206)	T, Tr, C
	Motherwort (p.156)	T, Tr, C
Heavy bleeding	Raspberry leaf (p.191)	T, Tr, C
	Nettle (pp.234–235)	T, Tr, C
	Yarrow (p.54)	T, Tr, C

WOMEN'S HEALTH PROBLEMS (CONTINUED)

CONDITION	HERB	PREPARATION
Scanty bleeding	Chinese angelica (p.68)	D, Tr, C
	White peony (p.172)	T, Tr, C
	Black cohosh (pp.100–101)	Tr, C
	Motherwort (p.156)	T, Tr, C
Irregular cycle	Chaste berry (pp.242–243)	Tr, C
	White peony (p.172)	T, Tr, C
	Black cohosh (pp.100–101)	Tr, C
Helping fertility	Chaste berry (pp.242–243)	Tr, C
	Motherwort (p.156)	T, Tr, C
Sore breasts	Calendula (p.86)	T, Tr, C
	Red clover (pp.226–227)	T, Tr, C
	Echinacea (p.118–119)	T, Tr, C
Thrush	Cat's claw (p.232)	Tr, C
	Oregano (p.171)	T, Tr, C
	Tea tree (p.163)	L (pessary)
Menopausal problems	Black cohosh (pp.100–101)	Tr, C
	Sage (p.201)	T (cooled), Tr, C
	Alfalfa (p.162)	C, or as food
	Puncture vine (p.224)	Tr, C
	Liquorice (pp.140–141)	Tr, C
	Wild yam (pp.116–117)	Tr, C
Osteoporosis	Black cohosh (pp.100–101)	Tr, C
	Withania (p.248)	Tr, C
	Alfalfa (p.162)	T, C, or as food

CAUTIONS FOR WOMEN'S HEALTH PROBLEMS

Seek immediate professional advice for:
• Persistent pain in abdomen or pelvis
• Vaginal bleeding between periods, after sex, or following the menopause
• Any unusual vaginal discharge
• Thickening, lump, or change in shape in a breast
• Discharge from a nipple

PREGNANCY

CONDITION	HERB	PREPARATION
Morning sickness	Ginger (pp.254–255)	T, C
	Chamomile (pp.98–99)	T, C
	Slippery elm (p.231)	T
Constipation	Flaxseed (pp.158–159)	Food
	Dandelion root (p.216)	D, C
	Senna (p.87)	C
Varicose veins/ haemorrhoids	Witch hazel (p.144)	L (distilled water)
Colds, flu colds	Elderflower/berry (pp.204–205)	T, S, C
	Sea buckthorn (p.146)	T, S, C
	Echinacea (pp.118–119)	T, C
	Plantain (p.181)	T, C
Poor sleep	Passion flower (p.173)	T, C
	Valerian (p.237)	T, C
	Skullcap (p.206)	T, C
	Lemon balm (p.164)	T

CAUTIONS FOR SKIN PROBLEMS

Seek immediate professional advice for:
- Prolonged nausea causing inability to eat
- Frequent vomiting
- Frequent urination for more than 3 days
- Breast pain with swollen glands under the arms or a fever
- Fluid retention that has not reduced after 3 days

See also: Pregnancy and after, pp.46–47

MEN'S HEALTH PROBLEMS

CONDITION	HERB	PREPARATION
Erectile dysfunction	Ginseng (pp.176–177)	Tr, C
	Ginkgo (pp.138–139)	Tr, C
	Saw palmetto (p.207)	C
	Puncture vine (p.224)	Tr, C
Infertility	Pumpkin seeds (p.109)	Food
	Ginseng (pp.176–177)	Tr, C
	Golden root (p.186)	Tr, C

MEN'S HEALTH PROBLEMS (CONTINUED)

CONDITION	HERB	PREPARATION
Prostate problems	Saw palmetto (p.207)	C
	Nettle root (pp.234–235)	C
	Pumpkin seeds (p.109)	Food

CAUTIONS FOR MEN'S HEALTH PROBLEMS

Seek immediate professional advice for:
• Swelling or lump in the testicle
• Change in shape or size of the testicle
• Total and persistent failure to get an erection

MENTAL AND EMOTIONAL PROBLEMS

CONDITION	HERB	PREPARATION
Anxiety/nervousness	Valerian (p.237)	D, Tr, C
	Lemon balm (p.164)	T, Tr, C
	Limeflowers (pp.222–223)	T, Tr, C
	Motherwort (p.156)	T, Tr, C
Depressed mood	St John's wort (pp.148–149)	Tr, C
	Damiana (p.230)	T, Tr, C
	Golden root (p.186)	Tr, C
	Rosemary (p.190)	T, Tr, C
Chronic stress	Siberian ginseng (pp.124–125)	Tr, C
	Ginseng (pp.176–177)	Tr, C
	Withania (p.248)	Tr, C
	Liquorice (pp.140–141)	T, Tr, C
SAD (Winter "blues")	St John's wort (pp.148–149)	Tr, C
	Golden root (p.186)	Tr, C
	Rosemary (p.190)	T, Tr, C
Difficulty in sleeping	Passion flower (p.173)	T, Tr, C
	Valerian (p.237)	D, Tr, C
	Californian poppy (p.128)	T, Tr, C
	Hops (p.146)	Tr, C
Nervous exhaustion	Skullcap (p.206)	T, Tr, C
	Vervain (p.240)	T, Tr, C
	Oats (p.74)	T, Tr, C
	Siberian ginseng (pp.124–125)	Tr, C

MENTAL AND EMOTIONAL PROBLEMS (CONTINUED)

CONDITION	HERB	PREPARATION
Poor memory and concentration / failing memory	Dan shen (p.197)	D, Tr, C
	Ginkgo (pp.138–139)	Tr, C
	Lemon balm (p.164)	T, Tr, C
	Rosemary (p.190)	T, Tr, C
	Sage (p.201)	T, Tr, C
	Schisandra (p.202)	T, Tr, C

CAUTIONS FOR MENTAL AND EMOTIONAL PROBLEMS

Seek professional help and advice for persistent or severe emotional and nervous problems.

PROMOTING HEALTH AND PERFORMANCE

CONDITION	HERB	PREPARATION
Physical stamina	Ginseng (pp.176–177)	Tr, C
	Siberian ginseng (pp.124–125)	Tr, C
	Golden root (p.186)	Tr, C
	Ginkgo (pp.138–139)	Tr, C
Exams	Rosemary (p.190)	T, Tr, C
	Ginkgo (pp.138–139)	Tr, C
	Siberian ginseng (pp.124–125)	Tr, C
	Schisandra (p.202)	T, Tr, C

SUPPORTING IMMUNE FUNCTION

CONDITION	HERB	PREPARATION
Chronic infection	Astragalus (p.73)	Tr, C
	Echinacea (pp.118–119)	Tr, C
	Cat's claw (p.232)	Tr, C
	Siberian ginseng (pp.124–125)	Tr, C
Depleted immune system	Reishi (p.135)	C
	Shiitake (p.151)	C, or as food
	Astragalus (p.73)	Tr, C
	Golden root (p.186)	Tr, C

Children's common health problems

1. The following are recommendations specific for the home treatment of children between 6 months and 12 years of age. The herbal remedies selected are considered safe for children between these ages. For babies up to 6 months, seek professional advice before giving them a herbal remedy.
2. Common health problems, as listed below, can be safely treated with herbal remedies. For other conditions, seek advice from your herbal or medical practitioner.
3. If your child is showing any of the signs listed under the Cautions, seek immediate medical advice and treatment. In children, potentially serious illness needs to be treated as quickly as possible. If in doubt, always err on the side of caution and seek advice.

DOSAGE LEVELS FOR CHILDREN

NB Do not give babies under 6 months herbal remedies without professional advice. You may need to adjust dosage levels for children who are particularly small or large for their age.

From 6 months to 1 year
1/10 minimum adult dose

From 1 to 6 years
1/3 minimum adult dose

From 7 to 11 years
½ minimum adult dose

From 12 to 16 years
Low adult dose

See also general advice on pp.44–45 and on p.47.

HEAD AND CHEST

CONDITION	HERB	PREPARATION
Tension headache and migraine	Limeflower (pp.222–223)	T, Tr, C
	Lemon balm (p.164)	T, Tr, C
	Lavender (pp.152–153)	L (diluted EO)
Earache	Garlic (pp.60–61)	L (O), C
	Elderflower (pp.204–205)	T, Tr, C, S
	Lavender (pp.152–153)	L (diluted EO)
Styes	Calendula (p.86)	L (ointment)
	Golden seal (p.147)	L (ointment)
Conjunctivitis	Eyebright (p.129)	L (lotion), T, Tr, C
	Echinacea (pp.118–119)	T, Tr, C
	Witch hazel (p.144)	L (distilled water)
Sore eyes/lids	Witch hazel (p.144)	L (distilled water)
	Chamomile (pp.98–99)	L (T)
Toothache	Clove (p.123)	L (1 clove or 1 drop EO)

HEAD AND CHEST (CONTINUED)

CONDITION	HERB	PREPARATION
Mouth ulcers/oral thrush	Liquorice (pp.140–141)	L, (neat, Tr)
	Echinacea (pp.118–119)	L (diluted Tr)
Nosebleed	Nettle (pp.234–235)	T, Tr, C
Hay fever and allergy	Eyebright (p.129)	T, Tr, C
	Baical skullcap (p.203)	T, Tr, C
	Elderflower (pp.204–205)	T, Tr, C
Feverish states	Elderflower (p.204–205)	T
	Meadowsweet (p.130)	T
	Limeflowers (p.222–223)	T
Colds, flu colds	Elderflower/berry (pp.204–205)	T, Tr, S
	Echinacea (pp.118–119)	T, Tr, C
	Cinnamon (p.97)	T, Tr, C
	Thyme (p.221)	T, Tr, C, S
Sore throat	Echinacea (pp.118–119)	G T, Tr
	Plantain (p.181)	G T, Tr
	Liquorice (pp.140–141)	G T, Tr
Cough, catarrh, and congestion	Elderflower (pp.204–205)	T, Tr, C
	Thyme (p.221)	T, Tr, C
	Garlic (pp.60–61)	C, or with food
Preventing infection	Elderberry (pp.204–205)	T, Tr, S
	Sea buckthorn (p.146)	S
	Blackcurrant (p.187)	Juice, S

CAUTIONS FOR HEAD AND CHEST CONDITIONS

Seek immediate professional advice for:
- Fever of 39°C (102°F) or above
- Heavy nosebleed lasting more than 1 hour
- Persistent one-sided headache or headache or pain that fails to improve within 48 hours despite over-the-counter medication
- Double vision/visual disturbance
- Unexplained dizziness
- Sudden or severe allergy
- Allergic reactions, including asthma
- Allergies that worsen after taking herbal remedies
- Persistent hoarseness, cough, or sore throat
- Chest pain or shortness of breath
- Coughing up blood

DIGESTION

CONDITION	HERB	PREPARATION
Poor appetite	Dandelion root (p.216)	Tr
	Alfalfa (p.162)	T, or as food
Stomach ache	Meadowsweet (p.130)	T, T
	Marshmallow (p.66)	T, Tr
	Chamomile (pp.98–99)	T, Tr
Nausea/motion sickness	Ginger (pp.254–255)	T, C
	Chamomile (pp.98–99)	T, Tr, C
Indigestion, wind, and bloating	Fennel (p.131)	T, Tr, C
	Caraway (p.94)	T, Tr, C
Digestive infections	Garlic (pp.60–61)	C, or with food
	Echinacea (pp.118–119)	Tr, C
	Cinnamon (p.97)	T, Tr, C
Worms	Pumpkin seeds (p.109)	Ground seeds
	Thyme (p.221)	T, Tr, C
	Garlic (pp.60–61)	C, or with food
Cramps and pain	Cramp bark (p.241)	T, Tr, C
	Chamomile (pp.98–99)	T, Tr, C
	Ginger (pp.254–255)	T, C
Diarrhoea	Plantain (p.181)	T, Tr, C
	Raspberry leaf (p.191)	T, Tr, C
	Slippery elm (p.231)	T, C
Constipation	Flaxseed (pp.158–159)	T
	Slippery elm (p.231)	T

CAUTIONS FOR DIGESTIVE CONDITIONS

Seek immediate professional advice for:
- Difficulty in swallowing
- Persistent abdominal pain or indigestion
- Passing blood in the stools
- Weight loss or failure to thrive

Glossary

A

Adaptogen Aids the body in adapting to stress, supports healthy function

Aerial parts Above-ground parts of the plant

Analgesic Reduces or relieves pain

Antibacterial Combats bacterial infection

Antibiotic Destroys or inhibits micro-organisms

Anti-catarrhal Reduces or relieves catarrh and congestion

Anticoagulant Prevents blood clotting, thins the blood

Anti-emetic Reduces or relieves nausea and sickness

Anti-fungal Combats fungal infection

Anti-haemorrhagic Reduces or stops bleeding

Anti-inflammatory Reduces inflammation

Antioxidant Prevents oxidation and breakdown of tissues

Antiseptic Destroys or inhibits micro-organisms that cause infection

Antispasmodic Relieves muscle cramps or reduces muscle tone

Anti-viral Combats viral infection

Aphrodisiac Excites libido and the sexual organs

Aromatic Having an aroma

Astringent Tightens mucous membranes and skin, reducing secretions and bleeding from abrasions

Autoimmune Acute or chronic illness caused by immune system attacking itself

Ayurveda Traditional Indian and Sri Lankan system of medicine

B

Bitter Bitter taste stimulates flow of saliva and digestive juices, increasing appetite

C

Carminative Relieves digestive gas, bloating, and indigestion

Circulatory stimulant Stimulates blood flow, usually to a given area, e.g. hands and feet

CITES Convention on International Trade and Endangered Species

Compress A cloth pad soaked in hot or cold herbal extract and applied firmly to the skin

Counter-irritant Irritant to the skin used to relieve more deep-seated pain or discomfort

Cream A mixture of water with fat or oil that blends with the skin

D

Decoction Water-based preparation of root, bark, berries, or seeds simmered in boiling water

Demulcent Coats, soothes, and protects body surfaces such as the mucous membranes of the digestive tract

Depurative Detoxifying agent

Detoxification The process of aiding removal of toxins and waste products from the body, especially via liver and kidneys

Diaphoretic Induces sweating

Diuretic Stimulates urine flow

E

Emetic Causes vomiting

Emollient Softens or soothes the skin

Essential oil Aromatic oil distilled from plants containing volatile oils

Expectorant Stimulates more effective coughing and clearance of phlegm from the throat and chest

F

Fixed oil A non-volatile oil (plant constituent). An oil produced by hot or cold infusion (preparation)

I

Immune modulator Promotes coordinated response by the body's immune defences to counter infection and inflammation

Immune stimulant Stimulates the body's immune defences to counter infection

Immuno-suppressant Inhibits or blocks the body's normal immune defences

Infusion Water-based preparation in which flowers, leaves, or stems are brewed in a similar way to tea

Inhalation Breathing of medicinally infused steam or liquid through the nasal passages

Interactions Where a herb and drug are taken at the same time, changing the effect of the drug (or herb), or producing an adverse reaction

L

Latex Sticky white or yellow juice released when plant part is broken, e.g. Dandelion leaf (Taraxacum officinale)

Laxative Promotes evacuation of bowels

N

Nervine Restores the nerves, relaxes the nervous system

Neuralgia Pain caused by nerve irritation or damage

Nutritive Provides nutritional input

O

Oestrogenic Has oestrogen-type hormonal activity within the body

Ointment A blend of fats or oils that form a protective layer over the skin

P

Phytochemistry The study of plant chemistry – plant compounds and their make-up

Placebo A substance with no medicinal effect, used as a control in testing new medicines

Poultice Herbal preparation applied locally to alleviate pain or swelling

S

Sedative Reduces activity and nervous excitement

Spasmolytic Relaxes muscles

Standardized extract Herbal extract produced with defined level of key constituent(s)

Stimulant Increases rate of activity and nervous excitement

Synergy Where the combined effect, e.g. of a herbal remedy, is greater than the sum of the effects of its constituents

Synthetic Chemicals or medicines produced artificially in a laboratory rather than derived from natural products

Systemic Affecting the whole body

T

Tincture Liquid herbal preparation made by soaking herb in water and alcohol

Tonic Exerts a restorative or stimulant action on the body

Topical Application of herbal remedy to body surface

Tranquillizer Has relaxing and sedative properties

Tuber A swollen part of an underground stem, e.g. potato

V

Volatile oil Plant constituent distilled to produce essential oil

W

Wild crafting Gathering herb material from wild rather than cultivated plants

Resources

FINDING A HERBAL PRACTITIONER

Herbal advice and medicines from a caring and knowledgeable professional can be invaluable when you are looking for a natural approach to health problems. Finding a well-qualified and professionally registered medical herbalist is not always easy. Recommendations from family and friends are helpful but check that your practitioner is professionally registered. Members of the following associations are governed by a strict code of ethics and have all received intensive training in herbal and medical sciences. Most have BSc degrees in herbal medicine.

National Institute of Medical Herbalists
Clover House, James Court
South Street
Exeter EX1 1EE
www.nimh.org.uk
Largest and oldest professional association of medical herbalists in the UK, with members throughout the UK and worldwide. Useful links on website, including information on training courses.

College of Practitioners of Phytotherapy
Oak Glade
9 Hythe Close
Polegate
East Sussex BN26 6LQ
www. phytotherapists.org
Publishes the *British Journal of Phytotherapy*. Key scientific links on website.

Other UK professional associations registering herbal practitioners include:

Association of Master Herbalists
www.associationofmasterherbalists.co.uk

International Register of Consultant Herbalists
www.irch.org

United Register of Herbal Practitioners
www.urhp.com
All the above associations are members of the European Herbal and Traditional Medicine Practitioners Association (contact: www.ehtpa.eu), an "umbrella" organisation for practitioner associations within the European Union.

IRELAND
Irish Institute of Medical Herbalists
www.iimh.org

Irish Medical Herbalists Organisation
www.mkdesign.ie/imho.html
Both organisations provide lists of well-qualified herbal practitioners.

AUSTRALIA
National Herbalists Association of Australia (NHAA)
PO Box 696
Ashfield NSW 1800
www.nhaa.org.au
Australia's oldest natural therapies association and only national professional body of medical herbalists. Website offers information on training courses and seminars, provides a herbal medicines discussion board, and allows you to search for a herbalist in your area.

NEW ZEALAND
New Zealand Association of Medical Herbalists (NZAMH)
nzamh.org.nz
PO Box 12582
Chartwell, Hamilton 3248
www.nzamh.org.nz
New Zealand's organisation of professional medical herbalists. Read the latest articles about herbal medicines or find a herbalist in your area.

HERBAL MEDICINE SUPPLIERS
UK
G. Baldwin & Co
171/173 Walworth Road
London SE17 1RW
www.baldwins.co.uk
Oldest herbal suppliers in London, stocking wide range of herbs and tinctures, also mail order.

Neal's Yard Remedies
8–10 Neal's Yard
Covent Garden
London WC2H 9DP
www.nealsyardremedies.com
Good range of organic herbs and tinctures supplied mail order or in shops.

Napiers
18 Bristo Place
Edinburgh EH1 1EZ
www.napiers.net
Herbal advice line staffed by qualified herbalists: 0131 225 5542
Wide range of herbs and tinctures stocked in shops or mail order.

IRELAND
Clareherbs
9 Sea Road
Co. Galway
www.drclare.net

AUSTRALIA
Austral Herbs
02 6778 7357
www.australherbs.com.au
Comprehensive online shop, supplying dried herbs, spices, and botanicals to customers worldwide.

NEW ZEALAND
New Zealand Herbals
26 Conway Street
Christchurch 8024
www.nzherbal.co.nz
Provides natural health-care services and products, with an extensive range of herbal supplements. Purchase herbal remedies, creams, and ointments online.

HERBAL INTEREST
The Herb Society
PO Box 196
Liverpool L18 1XE
www.herbsociety.org.uk
The Herb Society aims to promote interest in all aspects of herbal medicine – medicinal, culinary and horticultural. It runs a national herb garden in Sulgrave, Northamptonshire, plus local groups, meetings, and conferences. Website has useful information, discussion board, and good links. It also lists other herb societies worldwide. A key organisation

for everyone interested in herbal medicine, it also publishes *Herbs*, a first-rate magazine.

TRAINING COURSES
Discovering Herbal Medicine
www.newvitality.org.uk
A one-year correspondence course suitable for anyone interested in learning more about herbs and how to use them safely and confidently in their daily life. The course has been running for over 30 years and has been regularly updated. Highly recommended.

PROFESSIONAL TRAINING COURSES
UK: A BSc (Hons) course in Herbal Medicine now runs at Westminster University in London. An MSc in Herbal Medicine is offered by the University of Central Lancashire, for healthcare practitioners who want to also train as medical herbalists. A distance learning programme run by the National Institute of Medical Herbalists has recently started and will lead to membership of this professional association, though not to a recognised degree.

OTHER WEBSITES AND CONTACTS
American Botanical Council
www.abc.herbalgram.org
A key international resource, with lots of useful information and links. Publishes *Herbalgram*, perhaps the most informative English-language publication for those interested in herbal medicine.

Chelsea Physic Garden
66 Royal Hospital Road
London SW3 4HS
www.chelseaphysicgarden.co.uk
London's oldest medicinal herb garden in the middle of Chelsea.

Henriette's Herbal Homepage
www.henriettes-herb.com
Culinary and medicinal herb site

Medline
www.nlm.nih.gov
Essential database for accessing scientific papers on herbal medicine.

Index

Acknowledgments

About the Author

Andrew Chevallier is an experienced medical herbalist who has been in private practice in London and Norfolk, UK for over 30 years. A leading figure in the field of herbal medicine, he is a past President of the National Institute of Medical Herbalists and former vice-Chairman of the European Herbal Practitioners Association. He was a Senior Lecturer in Herbal Medicine at Middlesex University for 10 years, having helped to develop its BSc (Hons) degree course in Herbal Medicine, the first of its kind at European university. His books include *Encyclopedia of Herbal Medicine* and *Herbal Medicine for the Menopause*.

Author's acknowledgments

First edition

With many thanks to Anne Stobart and Rowan Hamilton for reading the first draft and suggesting key improvements; to Dr David Keifer for his astute recommendations on cautions; to Maria Chevallier for her active support and good humour; and last but not least, to the staff at DK, who were a pleasure to work with, displayed total professionalism, and made so many positive improvements to the look and feel of the book.

This edition

With many thanks to both David Hoffman, who consulted on the US edition, and saved me from several botanical howlers, and to all the staff at DK, especially to Kathryn Meeker, who were invariably full of good advice, always cheerful, and quick to save me from a host of unintended errors.

Publisher's acknowledgments

Dorling Kindersley would like to thank photographer Sian Irvine and her assistant Byll Pullman; illustrators Karen Gavin, Gillie Newman, and Ryn Frank; Diana Vowles, Hilary Mandleberg, Andrea Bagg, and Tara Woolnough for editorial assistance; Sara Noonan for design assistance; and Lynn Bresler for the original index. Thanks to Dr David Keifer for reviewing the US edition, and to Jane Daley for reviewing the Australian edition.

First edition

DK INDIA: Designers Arunesh Talapatra, Tannishtha Chakraborty **Editors** Dipali Singh, Aditi Ray, Pankhoori Sinha **DTP Designers** Sunil Sharma, Harish Aggarwal, Govind Mittal, Pushpak Tyagi **DK UK: Senior Editor** Jennifer Latham **Managing Art Editor** Marianne Markham **Managing Editor** Penny Warren

Picture Research Myriam Megharbi, Romaine Werblow, Claire Bowers, Lucy Claxton **Production Controller** Rebecca Short